Spencer Gerhardt

Ticking Stripe

Blank Forms Editions

Antarctica Starts Here
5

I

As a young musician growing up in the Midwest of the 1990s, I was drawn to a certain kind of 1960s "world-building" art, especially of the minimal, conceptual, and New York varieties, which moved across genres through a general and immersive aesthetic, and embodied a spirit of authentic subculture endemic to the nineties. This encompassing perspective was most broadly recognized through Andy Warhol and the Velvet Underground, but also in an initial, unadulterated form by La Monte Young and Marian Zazeela, whose mythic and expansionary sound-and-light constructions had projected from a Church Street loft uninterrupted since the 1960s.

At the same time, largely forgotten figures of the era were being rediscovered—people like Tony Conrad, Jack Smith, and Henry Flynt, who worked in the same downtown New York circles as Young, Zazeela, Warhol, Yoko Ono, Simone Forti, Walter De Maria, and others, but were deemed too difficult or unconventional by the standards of the time, or whose version of expanded art stretched beyond recognizable and well-ordered figure into the flattening horizon of unclassifiability. To a suburban teenager, this lofty time and setting held out the promise of a true artistic environment: full of life and ideas, exotic and intransigent figures, and rich but elusive work, carried out with immediacy in larger and often transposable contexts.

In terms of my work, I was drawn to an idea of minimalism that combined idealized or romantic introspective states with some ambient (and perhaps nonstandard) foundational perspectives, where music flowed naturally out of a general and immersive artistic environment. Large parts of this book could be seen as attempts to clarify this idea, through the creations of this first generation of minimalist artists who I was fortunate to get to know, and the work of likeminded friends. Indeed, while the title is intended to evoke a construction of the continuum and, in less abstract terms, a common pattern on synthetic pillows, it could also describe a process of creation, carried out in evolving and expanding contexts through conversations with artists, musicians, and mathematicians.

In broad terms, the book pairs notions of construction and continuity, evidenced in math through figures like L. E. J. Brouwer but also in the initial strands of musical minimalism, and through this distinguished lineage much subsequent work in visual art, composition, and popular music (Brian Eno famously characterized Young as "the daddy of us all"). The included writings attempt to develop an aesthetic framework for this pairing, captured in various lights and forms. In some cases, this is expressed directly through the foundational ideas that surrounded the development of minimalism out of modernist sources. In others, it appears indirectly by way of facets of a construction or formal method, or features in a painting. While much of the writing describes a general setting, like the depiction of coiled rind in Dutch still life, a few technical examples from logic and algebra are included, adding regions of greater detail within an overall picture. There are also "figures" provided by friends, whose work is not always explicitly described, although it runs between the lines of this collection. Taken as a whole, the book develops a setting for a certain kind of nonstandard romantic and introspective music, and provides one interpretation for a sizable body of work that has emerged out of the initial strands of musical minimalism.

II

As an undergraduate at the University of Wisconsin, through some generous resources available to students, I organized a film and music series with the idea of inviting out interesting artists. One day I tried calling Tony Conrad's office number at the University of Buffalo, which I located on a primitive university directory website. Much to my surprise, Conrad picked up and, without evident hesitation, agreed to come out to Madison for a few days to perform some of his classic early minimalist music and show a selection of his films and video art.

Conrad did attract considerable attention in the Midwest of the late 1990s for what might be characterized as an inscrutable personal style. I remember in particular a straight-legged, lime-green polyester suit under a canonical garment of the era: the overstuffed yellow-and-black North Face puffer. Based on such expressive choices, which seemed to straddle a line between humor, performance, and provocation familiar to his 1970s video art, we were sent drinks at various collegiate bars and admonished to enter a frat party on the way back to his hotel. By this time quite agreeable, Conrad walked through the door, looked around, and producing his signature expression of charmed amazement, exclaimed, "Well, I haven't been to a party like this in ten years!"

Like Stendhal before him, Conrad initially set out to be a mathematician before getting fully sidetracked by the 1960s. As an enthusiastic young logic student and musician, I was an ideal sounding board for some of this less familiar history. In a monologue on music, math, art, and philosophy stretching across several days and every secondhand bookstore in Madison, Conrad recounted his years as a math student at Harvard in the late 1950s and his experience entering the untrammeled atmosphere of the early 1960s downtown New York art scene. From that formative training followed an almost inconceivable list of activities: performing with Young, Zazeela, John Cale, and Angus MacLise in the Theatre of Eternal Music; making some of the most enduring films of the sixties underground; recording *Outside the Dream Syndicate* (Caroline, 1973) with Faust and a grocery-store instructional dance single ("Do the Ostrich") with Lou Reed and Walter De Maria; creating video art with Mike Kelley and Tony Oursler; naming the Velvet Underground; and much more.

Conrad made reference to other "far out" figures of this milieu who came from math and philosophy, people like Henry Flynt, Dennis Johnson, and Catherine Christer Hennix. Flynt was a fellow math student at Harvard similarly drawn to the beacon of New York and Young's artistic circle. In the early 1960s, he introduced concept art, an important precursor to conceptual art explicitly drawn from certain foundational ideas in math. Johnson, a part of Young's circle since their undergraduate years at UCLA, was one of the few contemporaries that Young would cite as an influence. Deemed a romantic even by the loftiest standards of the era, Johnson wrote some of the earliest examples of minimalist music and also did significant work in topology and geometric group theory, eventually retreating completely into the California mountains. Similarly, Hennix started out as a logic, linguistics, and philosophy student in Sweden before hearing the siren call of Young's New York milieu. She subsequently took Young and Zazeela's world-building approach to minimalism into highly abstruse and conceptual territories.

Conrad recounted many funny stories, including his and Flynt's attempts to pass W. V. Quine's logic class ("the old Navy man") and the infamous "Was Greek Mathematics Crazy?" panel that he, Flynt, and Hennix participated in at Glenn Branca's loft (along with a young John Baez and Remko Scha), which led to additional long-lasting feuds in the world of minimalism. I remember being very intrigued by this unlikely intersection of math and the avant-garde, and by these figures whom Conrad himself might classify as "far out." Indeed, this setting seemed to constitute a unique and unrecognized scene in its own right, rife with activity and strained interpersonal relationships, periodically converging on areas of common interest, all occurring in close proximity to the canonical artists of the era but with little public presence or recognition.

III

At the time, I was taking a class proving Gödel's incompleteness theorems and locating them within certain historical foundational positions in math like logicism, formalism, and intuitionism. I was struck by the fact that at the turn of twentieth century, lively and often personal foundational disputes were taking place between the leading mathematicians of the era, perhaps akin to more familiar debates happening within the arts, and that each of these foundational views could also be taken to represent a certain kind of aesthetic standpoint.

From this perspective, I found something particularly intriguing about Brouwer's intuitionism. A pioneer of modern topology and one of the most important mathematicians of the first half of the twentieth century, Brouwer was commonly presented as the last-place finisher in these foundational disputes, offering a view that was perhaps ideologically sound but that few people liked and even fewer practiced. (David Hilbert likened it to "proscribing a boxer the use of his fists."[1]) While invariably expressed in terms of the rules available to do math, reading between the lines, it seemed to flow from a broader (and perhaps even less palatable) project: a reconstruction of the continuum according to how it is given in intuition.

While the composition of the continuum is one of the most vexing problems in the totality of human thought—Leibniz classified it as one of the two labyrinths of the human mind, along with the nature of free will—at the beginning of the twentieth century a standard mathematical construction existed, which in turn served as the default models for space and time. Brouwer felt this standard model had little correspondence with how the continuum was given intuitively, and of our experiences of time and space. In response, he developed an entirely new and completely rigorous mathematical theory based on a conviction that the continuum never reduces to atomistic pieces. This creation was very different and indeed incompatible with standard mathematical practice. In Brouwer's hands, the continuum was given as an incompletable subject-dependent construction, developing in time and always subject to further refinement.

To a young person, Brouwer's composition of the continuum and his introduction of the subject into the formalization of math seemed to be the romantic and conceptual project par excellence, superseding even the 1960s in world-building ambition, conceptual clarity, and technical skill. Perhaps because of its obvious aesthetic qualities and seeming lack of necessity, it appeared to me like carrying out math from the perspective of art—or, in light of 1960s activities, like art itself—but done better than related examples. Indeed, this construction seemed to clarify the ideas I was drawn to in minimalism by locating continued introspective processes within a framework of irreducible notions of continuity. It offered an ambient setting for interpreting key theoretical ideas in Young's music, like tuning as a function of time, and more broadly suggested a general foundational approach to minimalism. At the time, this seemed like a highly exciting individual project, very different from what other people were doing or thinking about in the late 1990s, although I later discovered that Hennix had already sensed such implicit connections in her vast expansion of the minimalist framework in the 1970s. These ideas, however, were never fully worked out, which is the basis of the essay "Minimalism and Foundations."

IV

Brouwer taught at the University of Amsterdam, where he was succeeded by his students Arend Heyting and Evert Willem Beth. They established a tradition of math and logic in Brouwer's spirit at the university. After finishing my undergraduate degree, I moved to Amsterdam to attend graduate school at the University of Amsterdam's Institute for Logic, Language and Computation, a leading research center for logic in the intersection of math, computer science, and linguistics.

The earliest piece in this collection, "A Construction Method for Modal Logics of Space," is a product of that time. From a technical standpoint, I prove the strong completeness of some topological and spatiotemporal logics for several topological (and ordered topological) spaces. Topology studies continuity or cohesion in a general sense, without committing a priori to a single kind of ambient space. By focusing in on the notions of continuity and continuous deformation while remaining agnostic about other features, topological spaces themselves are very general and coarse spatial structures, with no intrinsic concept of distance. This flexible approach to spatial structure and continuity appealed to me.

While the paper slightly extended existing completeness results, for the present purposes the most salient feature is the construction itself. Rather than using the typical abstract model-theoretic reasoning, in the Brouwerian spirit I construct the desired mathematical objects directly in an infinite series of explicitly described stages. As points are added in the construction, an infinite set of formulas are associated with each new point. These formulas are intended to be *exactly* the statements expressible in the given language that are true at that point once the construction is completed. Since every stage in the construction is finite, while the eventual structures are infinite and dense, the key step is proving that the purported flawless balance between syntax and semantics set in motion by the construction actually holds on the eventual model.

I had the good fortune of learning this syntactic approach from Dick de Jongh, who used such constructions to prove completeness results for temporal logics. A student of Heyting and Stephen Kleene and a leading figure in the field of intuitionistic logic, de Jongh fit within the Brouwerian tradition at the Institute. He had the look of a 1970s American actor, wearing Levi's 501s with matching denim shirt and jacket, and with full, windblown gray hair giving the impression that he had just hopped off a bicycle. This outward appearance seemed to mask a sort of inward reticence. Widely renowned for his ability to give exquisite and minimal syntactic arguments, expressed at precisely the right level of detail to reveal what was actually going on, he would only produce these insights between long intervals of silence, delivering them just above of the threshold of human hearing. I remember running into a colleague leaving his office who likened their meeting to deciphering the sound of a flickering candle. To me, de Jongh's unusual presentation style and the incongruity between inner and outer appearances only heightened the presence and atmosphere surrounding his eventual refined statements, and impressed upon me the beauty of working syntactically when carried out in a sufficiently discerning way.[2]

One virtue of this direct approach is that it offers a lot of control over how formulas are satisfied on the eventual model. Indeed, you can dictate this in as much detail as your mind can conceive. I saw how to use this construction to establish the strong completeness of natural topological and spatiotemporal logics for the rational line. I felt that somehow I could leverage these facts to prove the completeness of a natural spatiotemporal logic on the classical continuum. This problem became a permanent fixture in my mind. For the relevant temporal logics, there was an existing method for extending completeness results for the rational line to the classical continuum. I spent countless hours trying to do this for the combined topological and temporal logic by subtly controlling how formulas were satisfied on the rationals and by trying to fully understand the interactions between the topological and temporal operators. A few times I thought I had a solution, staying up for days trying to work it out, but in the end it simply proved beyond my reach. In fact, it turned out some Russian logicians had already been working on this problem for a decade, and it took another ten years to establish the result using more abstract model-theoretic methods.[3]

There is a familiar truth in math that when you spend long enough thinking about an object or construction, it begins to take on an internal life of its own, filtered in some opaque way through intuition, subjective experience, and more general features. This is commonly offered as an explanation for mathematical intuition. To me, thinking about these continued syntactic constructions felt analogous to my interests in music and art, and indeed, after a certain point, just different facets of the same project. The aesthetics of these ongoing constructions with increasing levels of discernment felt similar to a way of composing I liked, through ongoing detection of the audible structure of the harmonic series. Further, the process of associating temporal and topological statements to points where they will hold once an infinite process has been completed seemed like a notable extension of the interplay of syntax and semantics in 1960s process art and conceptual art, in some ways exceeding it in clarity, precision, and conceptual rigor. Although I was never able to solve the problem itself, the attempt—like many things in life—proved fruitful from a more general viewpoint.

V

While at the Institute, I would sometimes meet Remko Scha for lunch to talk about music. By that time a distinguished computational linguist, Scha had been involved in the downtown New York music scene of Sonic Youth, Glenn Branca, and others in the early 1980s and made the classic minimal industrial record *Machine Guitars* (Kremlin, 1982); he was also an important figure in the Dutch sound art scene of the 1990s. Scha mentioned that Catherine Christer Hennix used to live in Amsterdam and take part in the logic community, although he wasn't sure whether she was still around, as no one had seen or heard from her in years. In any case, he said, she was now only interested in Lacan. Toward the end of my stay in Amsterdam, Scha wrote to say that Hennix was actually still in town and now had an office at the Institute one afternoon a week. We made plans to visit, but I decided to stop by early.

Dropping in late one afternoon, I discovered what would become a familiar scene: Hennix hurriedly trying to print a large number of physics papers before the office closed for the day. Introducing myself as someone at the Institute, she immediately seized upon the opportunity to present one of her pet ideas, a highly unconventional (and indeed, under standard interpretations, false) claim relating to Gödel's incompleteness theorems, evidenced by a handful of papers she was waving around. Because I was just a young person looking forward to meeting this interesting figure, I remember being slightly taken aback by her urgent and somewhat confrontational tone, which seemed to arise out of purely internal conditions. Indeed, this sense of urgency seemed to preclude the possibility of supplying the context necessary to frame her provocative statements, leaving me with a slightly unsettled feeling.

Conrad mentioned that Flynt and Hennix had continued thinking about formal methods long after he stopped considering such topics. (He expressed the opinion that his long structural film *Articulation of Boolean Algebras for Film Opticals* [1975] was something of a personal dead end, leading him to switch gears to video art in the late 1970s.) As he suggested, Hennix had worked continuously and largely on her own terms since the 1960s on a nonstandard and difficult-to-characterize amalgamation of music, logic, Eastern philosophy, and other topics. Although at the time she had not released any music commercially, she had, like Conrad, engaged in a highly impressive and perhaps even wider-ranging list of activities. At various times, Hennix worked closely with Young, Zazeela, Flynt, and Pandit Pran Nath; made the classic piece of 1970s minimalism *The Electric Harpsichord*; wrote poetry; produced a vast amount of theoretical writing surrounding her minimalist music, sculpture, and painting; studied and later worked with the well-known Russian dissident and logician Alexander Esenin-Volpin; worked with Marvin Minsky at MIT's Artificial Intelligence Laboratory; and more.

Hennix and I began to meet regularly to talk about music and logic, our conversations always filtered through her own unique and proprietary blend of seemingly disparate ideas and traditions. One principle not observable from the perspective of the contemporary art world but taken axiomatically in certain 1960s artistic circles was the notion that being "far out" could be taken as an indefeasible Cartesian virtue and indeed a hierarchical quality. Hennix herself would often employ such qualifications: for instance, Morton Feldman, "not far out," although she later expressed interest in pieces like *Patterns in a Chromatic Field* (1981). In this specific 1960s atmosphere and mentality, Hennix had long outpaced all competitors. When I first visited her apartment on De Wittenkade, she appeared mysteriously out of an expanse of hastily stacked boxes and physics papers, wearing a crinkled track suit and holding a pipe. Even before pleasantries could be exchanged, she launched into a several-hour disquisition on the cosmological interpretations of her new piece *Soliton(e) Star* (2003), which I eventually came to realize was comprised of a few rationally tuned sine tones playing quietly on loop in the background.

Gradually, her urgent tone modulated and stretched out into a very enjoyable and illuminating conversation lasting the next twenty years. Over the course of her life, Hennix always had extremely good intuition about what to draw from in her work. She searched constantly for new ideas and methods, for the most far-out approaches of the time, which led her to many novel observations and talking points. Especially in regard to music and art, she was the first person to offer a whole host of compelling interpretations of minimalism and conceptual art, in addition to her own very significant work in these areas. I came to realize that some of her unconventional logic ideas, often inspired by her work with Esenin-Volpin, were philosophically quite interesting, although perhaps not always worked out in exactly the ways that were claimed.

Given her implacable personality, the essentially internal nature of her investigations, and the relative vacuum in which she worked, Hennix was not the best at providing the context necessary to appreciate her unique thoughts. This is captured in her theoretical writings like *The Yellow Book* (1989), which is full of interesting ideas that are not really rendered in an externally digestible manner. For the majority of her life, this seemed to hinder any public reception of her work—until her late, very productive years working with Blank Forms. Through our conversations, I began to get a sense of one way of looking at her project, or at least the parts of it with which I had some familiarity. This is the basis of the essay "Domains of Variation." Hennix intended her theoretical writings, I think, to operate more in the spirit of Duchamp (whom she admired a lot), rather than as systematic exposition. She wanted readers to consider them and then respond by adding their own inner qualifications. Hopefully this is a project other people will take up in the future.

VI

After finishing the logic degree in 2004, I wanted to spend more time writing music. I moved to San Diego to do some work with Charles Curtis, the consummate performer and interpreter of Young's music. I describe this experience in the essay "Reverse Perspectives." I got a software job that wasn't overly consuming and lived within the California vernacular, in a large and successively expanded 1970s house/compound in a canyon, shared with interesting composers like Ming Tsao and Uday Krishnakumar.

This was a propitious time for reexamining Young's work. Young and Zazeela had recently completed *Just Charles & Cello in the Romantic Chord* (2002–3), a major four-hour cello-and-light piece for Curtis, which expanded upon ideas in Young's canonical work *The Well-Tuned Piano* (1964–). There was also renewed activity in an ongoing project of recontextualizing some of Young's early music. In particular, Young and Curtis worked intensively to develop a new performance version of Young's first major composition, *Trio for Strings* (1958–), typically viewed as the first piece of minimalism. This new version was to be performed in just intonation, the tuning context of Young's mature work, and in the proportions specified in the initial sketches for the piece. My essay "Trio for Strings" takes up this setting, tracing the development of Young's style through *Trio* and its subsequent elaborations. I wrote it to correspond with a planned commercial release of the piece, although that didn't end up occurring until Dia issued it in 2022. The essay has been updated to reflect some more recent developments.

There was also an interesting older generation of artists in and around San Diego—mostly affiliated with the University of California, San Diego, and generally accessible—people like Jean-Pierre Gorin, Eleanor and David Antin, Haim Steinbach, and Babette Mangolte. Through Curtis and this scene, I met Julia Dzwonkowski and Kye Potter. Dzwonkowski was an MFA student, and both she and Potter were good friends with Steinbach and had worked closely with Tony Conrad in Buffalo. They made a huge amount of work together, clearly inspired by 1960s conceptual art but with its own inviting and slyly contrarian quality; it reminded me a bit of Conrad in this respect.

We hit it off immediately and, with Curtis and his family, began to do and work on many things together, including an umbrella music-and-art project called World Egg, the art show *Beschäftigungen* at UCSD's Marcuse Gallery, various retreats to Arcosanti, and other projects. I wrote "Object Paintings" for Dzwonkowski's MFA show in the spirit of this collaborative environment. The essay places Dzwonkowski and Potter's paintings in dialogue with Henry Flynt's concept art through the idea of a "logic of involuntary associations." While it's safe to say that it did not reconstruct their initial thinking surrounding the work, they enjoyed this kind of interpretation.

VII

Dzwonkowski was running the visual arts lecture series at the time and, together with Curtis in the music department, decided to bring Flynt out to UCSD. Given my background, I was primarily drawn to Flynt's self-described 1970s "avant-garde hillbilly" songs and his writing about music, the 1960s art scene, and most of all concept art. Although Flynt didn't use any particularly difficult style, I found some of his writing nearly as challenging to interpret as Hennix's, perhaps due to the largely unfamiliar nature of the ideas and the overall project.

I came to realize that in concept art, Flynt used ideas he detected in Young's early 1960s work to criticize formalism in math—and thereby math more broadly—as well as movements in art and music like process art and total serialism. In his parlance, he devised the highly idiosyncratic project of "applying new music to metamathematics." While this project represented something of a psychical inverse to my more positive efforts, I was amazed that Flynt had already made all of these connections between art and logic in the early 1960s, drawing explicitly upon foundational ideas like formalism and Carnap's logical positivism. Furthermore, he did this in precisely the same artistic circle as Young, De Maria, and Robert Morris, years in advance of full-blown minimalism and conceptual art. In this sense, concept art seemed to serve as a precursor of things to come.

I was particularly drawn to work like *Concept Art Version of Mathematics System 3/26/61* (1961/63) and pieces related to the composition of the continuum. In *Concept Art*, Flynt introduced a "logic" of subjective experience through successive introspective acts called selection sequences. In pieces like *Each Point on this Line Is a Composition* (1961), he seemed to challenge classical notions of the continuum by requiring more explicit constructions for their realization. Although Flynt cited formalism as the inspiration for concept art, to me these works seemed similar in spirit to Brouwer's construction of the continuum through choice sequences, as ongoing acts of attention of an idealized subject.

I remember meeting Flynt after his UCSD talk on dignity, at a dinner incongruously held at the Rock Bottom brewpub in La Jolla. I asked him about Brouwer's construction, which he immediately dismissed out of hand because "nobody believed in it." I thought this was an interesting criticism. Flynt said he knew a little about Brouwer's work in college, but that it was viewed as marginal and wasn't something he was responding to or saw as particularly relevant to his project. Nonetheless, this started a conversation on music, logic, and any number of other topics that we have continued since, typically at the Japanese grocery store Sunrise Mart when I visit New York.

Flynt initially came from North Carolina and can have a charming and almost Huell Howserian tone when detailing his foundational challenges to modern civilization. Like Conrad and Hennix, he has produced a huge quantity of work in widely different areas like philosophy, economics, music, visual art, and art history. Most of this work has been done on his own terms, but at times it has intersected with canonical figures and events: playing with the Velvet Underground; receiving guitar lessons from Lou Reed; forming Henry Flynt and the Insurrections, with De Maria on drums; playing violin on Arthur Russell's dance record *Kiss Me Again* (Sire, 1978), with David Byrne on guitar; and making the only systematic documentation of Basquiat's early graffiti. When Flynt and I meet, I always have the strong suspicion that he has an entire talk planned out in advance, based on his most recent interests and findings. From this, I subtly try to detour him through other recollections of the 1960s, reflections on concept art or his music, and other interesting topics, many of which appear in this collection.

VIII

As a result of my writing on *Trio*, I had the opportunity to study raga in New York with La Monte Young and Marian Zazeela. They were childhood heroes and it was highly exciting to have the chance to learn from these great artists. I met them for the first time at one of Curtis's performances of *Patterns in a Chromatic Field* in SoHo. Not having a background in Indian music, I remember asking Young and Zazeela what I could do to prepare for the lessons. Young replied to the effect that there was nothing that could be done to prepare for them, which was not totally inaccurate.

Young is famous for his devotion to his projects and his indifference to everything else. I've always suspected that this is where his use of long durations came from: anything he likes he wants to last for a long time, and anything he doesn't, he does not want present at all. From an external perspective, this has the effect of making things appear to move very gradually. His and Zazeela's indifference to time as it is conventionally given extended in many directions beyond their work, including their development of a non-twenty-four-hour day and an alternative calendar system. Young was even known to take several hours to wash his hair. When I asked their disciple Jung Hee Choi about this, she told me that he just really enjoyed the experience of warm water. To me, this anecdote seemed to capture the aesthetic of his music. Early on, he discovered the sound he liked, and from that point forward lived according to Rilke's longing, dwelling amid the waves with no homeland in time. As someone who grew up dreaming of New York in the 1960s, I saw this as a great virtue. What might have been forty years on the busy SoHo street below could be only a few weeks within the drifting terrarium atmosphere of the Dream House.

I would get a call telling me when to come over, usually late at night, at times like 11 p.m. or 1 a.m. I remember walking up the narrow stairs to the second floor, taking off my shoes, and entering a large loft divided by paths of stacked boxes, packed with both amazing and more mundane accumulations since the 1960s, with incense burning several generations deep and *The Tamburas of Pandit Pran Nath* (1982) playing in the background. There was a small area by the entrance walled off by boxes, with pillows and a collection of tamburas on the floor. Although the cumulative effect drew the eye in all directions at once, my attention naturally gravitated toward a very attractive looking dehumidifier in the corner. This made some sense, since tamburas are noto-riously finicky to tune, until I realized it was De Maria's *Instrument for La Monte Young* (1966).

While Young has a reputation for being a formidable presence, I found that he could be quite inviting within the context of his interests and the specific artistic world he set up. Like Arnold Schoenberg before him, in spite of his fame and importance as a composer, he took teaching very seriously. While he's known for his expansionary ideas and sound, the lessons themselves were completely pragmatic and, even from the perspective of a mathematician, beautifully structured. They were full of well-thought-out exercises, assignments to practice, and astute observations about raga and composition more generally, sometimes interspersed with amazing offhand comments and historical recollections. It was the ideal of teaching, firmly grounded in the practical while reflecting a completely inspiring approach to the subject.

Together with Choi, the four of us would sing scales and exercises together, trying to locate partials in the complex sound of the tambura. Teaching through singing together has to be one of the most immediately given and primordial methods, lending it some intrinsic and profound quality. At first, some exercises seemed to proceed slowly. I remember momentarily feeling a slight impatience but quickly realizing that the point was not so much comprehension as continued recognition: tuning to the partials in the tambura in time with increasing discernment. This process generally reflects Young's approach, in which structure is manifested directly through sound and ongoing auditory experience. While more external, "architectural" approaches can also be interesting, even from a math standpoint, Young's intrinsic or local perspective seemed to offer deep results and always felt like the one to emulate.

At times, Young would extemporaneously produce an interesting pedagogical example. During our last lesson, we were discussing the beginning of a raga and he was trying to make a point about the scale. A relevant example occurred to him that, as usual, we slowly sang together, in this case the melody to "Somewhere Over the Rainbow." While this was just an offhand observation, and not notable to anyone else present, to me, within that specific setting and artistic environment, it summarized years of imagination and pointed toward the aesthetic ideal I had envisioned all along: a pure romantic space filtered through some rigorous (albeit nonstandard) conceptions.

IX

I moved back to California to teach math and philosophy at a small college in the San Fernando Valley. In 2007 Los Angeles, it was still conceivable to rent a cheap one-bedroom Victorian house in Lincoln Heights. The house had high vaulted ceilings, picture-glass windows, and the most likable and time-agnostic Home Depot renovation I've encountered, with new medium-pile maroon carpeting joined by satin gold transition strips to nondescript beige linoleum flooring. A fresh layer of buckling semigloss cream paint covered every inch of surface and detailing in the house, which was nearly empty except for a piano, mattress, and a few math books. I remember the artists Ricky Swallow and Lesley Vance visiting before a Dodgers game, and Ricky jokingly referring to it as Jandek's house. Taking in the surroundings, he suggested the album title *Beige Off Broadway*, which did seem evocative of a 1970s-inspired solo singer-songwriter record. At the time, I was writing songs on the piano, often sung in low register based on some raga exercises I would do with Young and Zazeela, and working on more classically minimalist pieces.

I regularly taught logic classes and continued to think a bit about the problem of extending spatiotemporal completeness to the classical continuum. In looking for a way forward, I found it instructive to return to the original sources, in this case John McKinsey and Alfred Tarski's 1944 paper "The Algebra of Topology." Although the authors didn't use these terms, theirs was one of the first papers to establish completeness results for modal logics with topological interpretations. However, their framework was much different than the one I had used and this started me thinking more about algebra, which like other things I got very into.

From a music perspective, there is something natural about a branch of algebra called group theory, which studies symmetries in a general sense, and through this perspective connects naturally to music and art. Indeed, some specializations of music theory make heavy use of group theoretic tools. However, this was not where my interest came from. While teaching, I had the fortune of learning about an application of group theory in arithmetic geometry, the Galois theory of curve coverings, from Katherine Stevenson. This was my first encounter with current research-level algebra, and I found it highly exciting. It seemed to offer a new and very different approach to resolving the classical problem lying at the heart of this book, the tension between the discrete and the continuous.

I was particularly drawn to the relationship between arguments given in "characteristic zero," which are often viewed over spaces with rich, classical geometric and topological structure, and the corresponding arguments given in "positive characteristic," which are viewed over much stranger and non-intuitable spaces coming from algebra, with at best a coarse topological structure and no obvious geometric meaning. Amazingly, some results are "characteristic independent" and apply over vastly different kinds of spaces: from the rich and familiar, like the classical continuum, to infinitely many other strange and looping "clock-arithmetic" spaces, with no coherent notion of distance or magnitude.

To me, algebraic groups represented the most exciting setting for this kind of exploration. These are groups defined by nice equations, from which they inherit a coarse topological structure. Given the way in which they are defined, algebraic groups can be interpreted over many kinds of spaces: finite, discrete, continuous, characteristic zero, and positive characteristic. This gives a uniform approach for studying symmetry while remaining somewhat agnostic about the nature of the underlying space, and provides a way of relating the solutions of equations viewed over different kinds of spaces.

Given any algebraic structure, one intrinsic question concerns generation: What of kinds elements will form the entire structure if allowed to interact freely? This is a natural question within algebra, and is often useful in applications. In many cases, problems from different areas of math can be reduced to this type of algebraic question. For instance, in the 1980s, Dennis Johnson studied such a group-theoretic generation problem in the context of knot theory. My paper "Topological Generation of Special Linear Groups" completely resolves a generation problem of this sort for one of the most important and natural algebraic groups, viewed over infinitely many different spaces of arbitrary characteristic. Indeed, the exact conditions determined are as minimal and natural as they possibly could be.

Another thing I found intriguing about algebraic groups was a collection of tools that allow you to transfer properties from algebraic groups, which have topological structure and therefore some cohesive qualities, to corresponding finite groups, which are discrete and difficult to handle in any uniform way. In "Topological Generation," I use this transfer property to establish a connection between topological generation in algebraic groups and random generation in corresponding finite groups. In "Order Three Normalizers of 2-Groups," I make even more detailed use of these tools to completely determine a structural property of finite almost-simple groups (these are building blocks of finite groups, a bit like how prime numbers are building blocks of positive integers) by determining a related property in the corresponding algebraic groups.

I had the great fortune to learn about group theory and algebraic groups from Bob Guralnick. As anyone in math knows, this is a little like learning still-life painting from Chardin or Morandi; it was an honor to learn from and later work with him on these great subjects. Having known countless artists and musicians over the years, I don't think I've ever met anyone who quite equaled Guralnick in terms of passion and devotion to their subject. He provides a very inspiring approach not only to mathematics but to carrying out one's life's work more generally.

Another theme found in this book is that there is something to be gained from working in a personal way, even if it can be a slower and at times halting approach. I discuss many examples of continuing a project over the decades without an audience or any immediate context for appreciation or response. Although this way of working seems quite foreign in our current moment, such a generative, internally directed, lifelong, and ultimately romantic approach has been a source of inspiration for me. It also provides another interpretation of the title *Ticking Stripe*, as a continuing, intrinsic pattern, analogous to an ongoing construction of the continuum or lifelong approach to one's work.

X

Given the long period of time this book covers, there are many people who should be recognized. Let me start by thanking Lawrence Kumpf for seeing fit to publish this collection in the most flattering of contexts, Blank Forms. No contemporary organization has done more to present music in a general art context or to document specific postwar music and art scenes like the one covered in this book. It's an honor to have some place in this venerable organization. I'd also like to thank Alec Mapes-Frances for his impeccable graphic design sense and Ciarán Finlayson and Dana Kopel for their editing suggestions and helpful insights. In addition, I'd like to thank my parents, Terese and Terry, and my sister and brother-in-law, Teena and Matt, for their continuous support, as well as Kate, Laura, Perry, and Vicky Mulleavy.

There are many people who appear in this book implicitly or explicitly, or who made parts of it possible. For their support, I would also like to thank Thomas Ankersmit, Johan van Benthem, Jung Hee Choi, Ryan Conder, Matt Connors, Charles Curtis and the entire Curtis family, Julia Dzwonkowski, Henry Flynt, Gillian Garcia, Robert Guralnick, Catherine Christer Hennix, Dick de Jongh, Ellen Kenney, Ruby Neri, Philip Ording, Sung-Hwa Park, Matt Paweski, Kye Potter, Katherine Stevenson, Ricky Swallow, Lesley Vance, and La Monte Young and Marian Zazeela.

1   David Hilbert, "Die Grundlagen der Mathematik,"
     Abhandlungen aus dem Seminar der Hamburgischen
     Universität 6 (December 1928): 65–85. English
     translation in Jean van Heijenoort, ed., *From Frege
     to Gödel: A Source Book in Mathematical Logic,
     1897–1931* (Cambridge, MA: Harvard University Press,
     1967), 464–479.

2   By way of contrast, I had the fortune of learning more
     indirect model-theoretic methods from the highly char-
     ismatic public intellectual Johan van Benthem. In meet-
     ings, he had an uncanny ability to express everything
     just incrementally beyond your level of understanding,
     leaving you with the feeling that something amazing
     had been revealed and you just needed to fit the pieces
     together.

3   It was eventually proved in Ian Hodkinson, "On the
     Priorean temporal logic with 'around now' over the real
     line," *Journal of Logic and Computation* 24, issue 5
     (October 2014): 1071–1110.

Matt Connors
Untitled, 2023
Colored pencil on paper
Courtesy of the artist and The Modern Institute

Introduction to Matt Connors: Gui(l)de

Hi Matt,

Thanks for your thoughtful message. It's very helpful. I was particularly intrigued by your statement that the simplest solutions to problems in your work were often the most irrational ones. Is there a specific example of this you have in mind?

It's interesting because in math simplicity is typically taken as a kind of synonym for rationality. When someone says simple, they don't usually mean that the problem or object of study is itself simple (the objects of math almost invariably seem quite strange and mysterious), but rather that somebody has discovered a path starting from simple principles leading to a complex conclusion, or, starting from some concrete situation, has ascended to a more abstract viewpoint, using only the most "rational" or "necessary" steps along the way. "Simplicity" is a series of well-placed guideposts leading from the things closest to us out into the unknown.

While this account is appealing, I was interested in the idea that there might be other kinds of simplicity in math, ones that might fall more in line with aesthetic considerations in art. It does seem true that in art a combination of simplicity and irrationality is preferable to something that is simple alone. Your idea of expediency interested me in this regard.

It reminded me a little of the status of examples in math. In some cases, they are denigrated as expedients for the thing itself (theorems), or as crutches for the mind to hold onto until it can walk on its own. A sentiment expressed by Sacks—"all examples of the universal, since they must of necessity be particular and so partake of the individual, are misleading."[1] There is only "the least misleading example" . . .[2]

However, apart from any use in proving theorems or developing intuition, examples also provide an experience, one which is at least partly aesthetic, and which is, in a sense, similar to our appreciation of art or nature. There is a mixture of the necessary (given these conditions, such and such must hold) with the unnecessary (all of the other properties that just happen to be true of the particular case), which is pleasant to apprehend, especially in cases where these elements are not easily reconcilable.

I guess it is a well-known trope in painting to view a painter's work as a series of examples (something like Mondrian or Morandi), but it does give some explanation to the power of expediency. The balance of the necessary and the merely conventional can be recognized, but not easily understood.

Anyway, I'm rambling now and have to catch the plane. Ellen and I would love to look at your studio when I get back and it would be fun to try to hammer out something for the press release. Thanks again for your response.[3]

Spencer

1   Gerald E. Sacks, *Saturated Model Theory* (Reading, MA: W. A. Benjamin, Inc.), 4.

2   Sacks, *Saturated Model Theory*, 4.

3   From email sent to Connors on April 2, 2013.

Julia Dzwonkowski and Kye Potter
Green Plastic Bear Mother and Child Coin Bank, 2024
Ink on paper
Courtesy of the artists

Object Paintings
31

CONSIDERING the proceeding of a Painters worke I have; a desire hath possessed mee to imitate him: He maketh choice of the most convenient place and middle of everie wall, there to place a picture, laboured with all his skill and sufficiencie; and all void places about it he filleth up with antike Boscage or Crotesko works; which are fantasticall pictures, having no grace, but in the variety and strangenesse of them. And what are these my compositions in truth, other than antike workes, and monstrous bodies, patched and hudled up together of divers members, without any certaine or well ordered figure, having neither order, dependencie, or proportion, but casuall and framed by chance?

Desinit in piscem mulier formosa supernè.

   —Horace, *Ars Poetica* 4

A woman faire for parts superior,
Ends in a fish for parts inferior.

   —Michel de Montaigne, "Of Friendship"

I

In the introduction to *The Foundations of Arithmetic*, Gottlob Frege states his famous Context Principle: "never to ask for the meaning of a word in isolation, but only in the context of a proposition. . . . If [this] principle is not observed, one is almost forced to take as the meanings of words mental pictures or acts of the individual mind . . ."[1]

Frege uses the Context Principle to explain how it is possible to refer to objects such as numbers, which we can neither perceive directly, nor indirectly by way of scientific apparatus.

He writes:

> That we can form no idea of its content is therefore no reason
> for denying all meaning to a word, or for excluding it from our
> vocabulary. We are indeed imposed on by the opposite view
> because we will, when asking for the meaning of a word, consider
> it in isolation, which leads us to accept an idea as the meaning.
> Accordingly, any word for which we can find no corresponding
> mental picture appears to have no content. But we ought always to
> keep before our eyes a complete proposition. Only in a proposition
> have the words really a meaning.[2]
>
> . . .
>
> Only by adhering to [the Context Principle] can we, as I
> believe, avoid a physical view of number without slipping into a
> psychological view of it.[3]

The Context Principle was adopted by Wittgenstein, among others, as a guiding principle of philosophical analysis. In the *Tractatus* it is noted that "only the proposition has sense; only in the context of a proposition has a name meaning"[4] and, later, "an expression has meaning only in a proposition."[5] Even after rejecting Frege's central thesis of numbers as objects, Wittgenstein maintains a version of the Context Principle in the *Blue Book* of 1934–35:

> And the same, of course, could be said of any proposition: Without
> a sense, or without the thought, a proposition would be an utterly
> dead and trivial thing. And further it seems clear that no adding of
> inorganic signs can make the proposition live. And the conclusion
> which one draws from this is that what must be added to the dead
> signs in order to make a live proposition is something immaterial,
> with properties different from all mere signs. But if we had to name
> anything which is the life of the sign, we should have to say that it
> was its *use*.[6]

Here a formalist viewpoint on the foundations of mathematics is criticized by invoking a version of the Context Principle: "only in the stream of life do words have their meaning."

II

The materials for Duchamp's readymades are the ordinary articles of life, objects with little intrinsic resonance. It is a testimony to the profound ordinariness of these items—a bicycle wheel, a stool, a shovel, a urinal—that even today they seem commonplace. While everyday objects may possess many senses or intensions, and hence a subsequent psychological richness or opacity, these objects are not expressive in this sense. The transparency of these objects may be attributed to the fact that their appearance closely reflects their use. As such, the objects require little by way of explanation. Under normal circumstances, the objects comprising the readymades are viewed with a degree of indifference. A new picture of the objects emerges when they are placed outside of their useful contexts.

III

The Object Paintings of Julia Dzwonkoski and Kye Potter depict items purchased in thrift stores or on eBay that have been photographed and then painted (via the pictures) in a photorealistic manner.

The Object Paintings portray the extra-ordinary articles of life. Unlike material for the readymades, the objects are neither visually indifferent, nor absent of good or bad taste. Their appearance does not reflect their use.

In most cases, it is not clear what the objects would be useful for. Some appear to be relics (*2×4 face*, 2007; *Golden Knuckleballer*, 2007) that had, at one time, a private significance, but now exist at a remove from this purpose. Others appear as inscrutable creations of the human mind (*Red Plastic Mass*, 2007; *Broken Nose Long Hat Ornament*, 2007; *Coil Pot Tower*, 2007).

Perhaps in recognition of these uncertain foundations, several titles put forward one or more speculative uses for the objects (*Pink Lady Spoonholder*, 2007; *TV Napkin Holder / Vermont TV Guide Holder*, 2007; *Green Plastic Bear Mother* and *Child Coin Bank*, 2007). In other cases, the titles provide a literal, descriptive account of the object at hand (*Golden Knuckleballer, Red Plastic Mass, Broken Nose Long Hat Ornament*), with the additional provision that, if separated from the images, we would have no idea whatsoever what the descriptions refer to.

In every case, the useful significance of the objects in their present context is incommensurate with the existential reverberation of the objects themselves. The objects neither display nor conceal their use, but assert their precedence over it, sometimes to comic effect.

The unifying feature of the objects is precisely their "lifelikeness." With a single exception, each of the objects resemble a person (in body or spirit), animal, or home. It is their living appearances that allow these objects to assert a significance apart from any perceived use. They have a psychological richness and opacity that Duchamp sought to avoid in his readymades.

The objects were not intended to end up in thrift stores or on eBay. Through neglect, chance, or some other unknown method, they ended up in the public sphere, outside the realm in which they were originally conceived. However, the objects appear to exist comfortably at a remove from their origins. Unlike old photographs or home movies, they do not ask the viewer to reconstruct their surrounding circumstances, or evoke an emotion based on the impossibility of reconstituting the past. On the one hand they appear sui generis; on the other (and perhaps as a result) they assert a life and meaning of their own, independent of the act of creation or the surrounding circumstances. If the readymades extended the possibility of knowing from a new perspective, the Object Paintings explore the creative possibilities afforded by not knowing.

IV

Although the Object Paintings depict inanimate vessels, it is the lifelike qualities of the objects that are of primary interest. In this sense, the paintings resemble portraiture more closely than still life. While classical still-life painting is based on the optical, or on apprehending a setting in all of its forms, portraiture is based on the psychological, or depicting the inner truth of a subject. For this reason, portraiture will often depict the subject in abstraction from its environment, or with only minimal surrounding elements, as occurs in the Object Paintings.

Consider the Object Paintings as portraits: When we ask what inner truths of the subjects are to be captured, we are forced to consider the properties these objects possess, as opposed to those that we ascribe to them. For instance, we see a human visage in the piece of lumber depicted in *2×4 face*, and in that sense view it as different from an ordinary object, but still we do not subscribe to the view that there is a psychology of wood to be captured. What is depicted are the human qualities we project on the objects. The paintings appear to us as portraits of the subconscious.

V

The subjects of these portraits are the thoughts that pass through the mind of the beholder, rather than the inanimate vessels depicted. This reversal is indicated by the size of the paintings. Within the universe of pictorial representation the objects are scaleless (they are presented in isolation against a blue screen), but with respect to the actual world they are scaled to human size. Thus the kitchen appliance in *Pink-Lady Spoonholder* is about as tall as a child, and the ursine matriarch of *Green Plastic Bear Mother and Child Coin Bank* is the height of a small bear. Through painting, the eye of the beholder becomes the frame of reference for the objects.

VI

When the above view is taken, the choice of painting the objects rather than displaying them becomes clear. Painting has the effect of dignifying the objects—of making them more "meaningful" than their mere physical embodiment—by presenting them in a classical mode. We view the existential or "human" traits of the objects (not actually present) as even more exemplary when presented within portraits; we are inclined to read even more into the objects.

If it is granted that the true subjects of the Object Paintings are not the items depicted, but rather the psychological underpinnings of thought, the choice of painting style also becomes clear. Superficially it could be asked: if the paintings are to be executed in a photorealistic style, why not just display the photographs (just as one might say, if it is the objects that are of interest, then why paint them)?

Our first response is similar to the one given above: we view a painting in a different manner than we do a photograph, even if the optical impressions are closely related. New photographs are viewed as transparent; they document reality. Paintings are not treated as documents; they assert a life and meaning of their own.

Second, even though photorealistic painting is not viewed as a transparent medium, photorealism does afford a transparency in the act of artistic production. It provides a method for the artist to depict objects without supplying an interpretation in the act of artistic production. The act of production is essentially local activity, while the unifying global "atlas" is provided externally. As a result, any meaning we ascribe to the Object Paintings as portraits is the result of the viewer's projection, rather than a conscious or unconscious statement on the part of the artists. The decision to paint the objects against a blue screen serves as both confirmation and possible objection to this claim (in the sense that this specific lack of setting suggests additional connotations).

VII

Because the artists chose not to display the objects along with the paintings, the images paired with the titles function as symbols for the missing objects. For Dzwonkoski and Potter, rendering the objects against a blue screen— which specifically refers to the lack of any setting—is a way of presenting such symbols apart from any context. On the one hand, blue is simply a color used for backdrops upon which pictures of reality are projected. On the other, it refers to a state of affairs, that by definition cannot exist in any physical setting, namely the absence of physical surroundings.

In this way, the Object Paintings appear in violation of the Context Principle— they contain symbols for the missing objects apart from any physical or linguistic surroundings. The Object Paintings function to ask the meaning of a name in isolation.

VIII

Frege writes that "if the Context Principle is not observed, one is almost forced to take as the meanings of words mental pictures or acts of the individual mind." If the meanings of words were in fact mental pictures, then we should naturally be limited to judgments concerning objects of which we can have distinct thoughts. However, as Frege points out, we are constantly reasoning in the opposite direction:

> Even so concrete a thing as the Earth we are unable to imagine as we know it to be; instead, we content ourselves with a ball of moderate size, which serves us as a symbol for the Earth, though we know quite well it is very different from it. Thus even although our idea often fails entirely to coincide with what we want, we still make judgments about an object such as the Earth with considerable certainty, even where its size is in point.
>
> Time and time again we are led by our thought beyond the scope of our imagination, without thereby forfeiting the support we need for our inferences. Even if, as seems to be the case, it is impossible for men such as we are to think without ideas, it is still possible for their connexion with what we are thinking of to be entirely superficial, arbitrary, and conventional.[7]

IX

The Object Paintings bring before the mind precisely the phenomena Frege sought to be rid of, ideas that attach themselves to thought, or pass involuntarily before the mind, but have no role in logical judgment—in Frege's terms, the "entirely superficial, arbitrary, and conventional."

The paintings depict objects and not propositions. Viewers must place the symbols within the context of complete propositions to determine their meanings. In a fashion similar to the readymades, this proposition is not recoverable from the artist's intentions (or, implicitly, from the objects themselves, or creators of the objects); it is an act of the individual mind. However, unlike the readymades, the Object Paintings are symbols with involuntary and almost universal psychological connotations attached to them. They bring before the mind sentimental thoughts and emotions that are in no way contained in the symbols themselves, but instead reveal the psychological residue of thought.

While another of Frege's guiding principles was "to always separate sharply the psychological from the logical," one might say there is an opposing principle at work in the Object Paintings, to observe the psychological from a logical standpoint.[8] Rather than advance a self-reflexive and delimiting examination of the medium, they explore the beneficial role of not-knowing in the creative process.

1   Gottlob Frege, *The Foundations of Arithmetic* (New York: Harper Torchbooks, 1960), xxii.

2   Frege, *The Foundations of Arithmetic*, §71.

3   Frege, *The Foundations of Arithmetic*, §116.

4   Ludwig Wittgenstein, *Tractatus Logico-Philosophicus* (London: Routledge & Kegan Paul, 1922), §3.3.

5   Wittgenstein, *Tractatus Logico-Philosophicus*, §3.314.

6   Ludwig Wittgenstein, *The Blue and Brown Books* (Malden, MA: Blackwell Publishing), 4.

7   Frege, *The Foundations of Arithmetic*, §60.

8   Frege, *The Foundations of Arithmetic*, xxii.

Lesley Vance
Untitled, 2017
Oil on linen
Courtesy of David Kordansky Gallery

Topological Generation of
Special Linear Groups

43

Abstract

Let $C_1, \ldots, C_e$ be noncentral conjugacy classes of the algebraic group $G = SL_n(k)$ defined over a sufficiently large field k, and let $\Omega := C_1 \times \ldots \times C_e$. This paper determines necessary and sufficient conditions for the existence of a tuple $(x_1, \ldots, x_e) \in \Omega$ such that $\langle x_1, \ldots, x_e \rangle$ is Zariski dense in G. As a consequence, a new result concerning generic stabilizers in linear representations of algebraic groups is proved, and existing results on random (r, s)-generation of finite groups of Lie type are strengthened.

Let $C_1, \ldots, C_e$ be conjugacy classes of a group G, and $\Omega := C_1 \times \ldots \times C_e$. The problem of specifying conditions for tuples $\omega \in \Omega$ which generate G, or which generate a subgroup of G with special properties, arises in many different contexts. For instance, if $C_1, \ldots, C_e$ are conjugacy classes of $G = SL_n(\mathbb{C})$, the Deligne-Simpson problem asks for necessary and sufficient conditions for the existence of a tuple $(x_1, \ldots, x_e) \in \Omega$ such that $x_1 \cdots x_e = 1$ and $\langle x_1, \ldots, x_e \rangle$ acts irreducibly on the natural module. A solution to this question yields information concerning monodromy groups of regular systems of differential equations on $\mathbb{C}P^1$ (see [42] for further details), and has been studied in numerous settings and generalizations [9] [36] [38] [39] [40] [41] [58]. In [10] and [11] necessary and sufficient conditions are given for the existence of such a tuple, although work on the many generalizations remains.

In a different direction, any finitely generated group which is generated by conjugates of a single element arises as a quotient of a knot group (see [31] for additional details), and such groups have been studied in a variety of contexts [8] [37] [13] [53] [60]. In the special case where G is a finite simple or almost simple group, the number of conjugates of a fixed element required to generate G is examined in [28] and [30].

Finally, a good deal of recent work concerns the notion of invariable generation [22] [21] [20] [33] [34] [61] [62], in which every tuple in a product of conjugacy classes generates G. In particular, the question of whether linear groups such as $G = SL_n(\mathbb{Q})$ or $G = SL_n(\mathbb{Z})$, $n \geq 3$, are invariably generated has attracted considerable attention.

In this paper we examine another class generation problem for linear algebraic groups of the form $G = SL_n(k)$, where k is initially taken to be an uncountable algebraically closed field of arbitrary characteristic. Once the desired result has been established, we then show the field conditions can be relaxed considerably (to arbitrary fields of characteristic zero, and algebraically closed fields of positive characteristic which are not algebraic over a finite field). Finally, we see the solution to this problem has interesting applications to generic stabilizers in linear representations of algebraic groups, and random generation of finite groups of Lie type.

Let us briefly describe the set up of the paper. Assume G is a simple algebraic group defined over an algebraically closed field k which is not algebraic over a finite field. Set $\Omega = C_1 \times \ldots \times C_e$ and say $(x_1, \ldots, x_e) \in \Omega$ *generates* G *topologically* if $\langle x_1, \ldots, x_e \rangle$ is Zariski dense in G. Since G is infinitely generated, the relevant finitary notion of generation is topological. In what follows, we determine necessary and sufficient conditions for the existence of a tuple $\omega \in \Omega$ which topologically generates $G = SL_n(k)$. In joint work with

Burness and Guralnick [5] [6], conditions for the remaining simple algebraic groups are determined.

Recall a conjugacy class C of G is called *quadratic* if a representative of C has a degree two minimal polynomial. Our main result is as follows.

Theorem 1.1. Let $C_1, \ldots, C_e$ be noncentral conjugacy classes of the algebraic group $G = SL_n(k)$ where k is an uncountable algebraically closed field, and $n \geqslant 3$. Let γ_i be the dimension of the largest eigenspace of a representative of C_i on the natural module for G. Then there is a tuple $\omega \in \Omega$ topologically generating G if and only if the following conditions hold:

(i) $\sum_{i=1}^{e} \gamma_i \leqslant n(e-1)$;
(ii) it is not the case that $e = 2$ and C_1, C_2 are quadratic.

Let us make a few comments about the theorem. First, the assumption that $n > 2$ is made for purposes of uniformity. **Theorem 4.5** below provides the relevant necessary and sufficient conditions for $SL_2(k)$. Second, it follows from the above theorem that for any noncentral conjugacy class C of $G = SL_n(k)$, $n \geqslant 3$, it is possible to generate G topologically by choosing n elements from C. Furthermore, this bound is sharp. If C is a class of transvections, it is easy to see that at least n conjugates in C are required to generate G topologically.

The notion of a generic property underlies much of the reasoning given below, so let us fix its meaning in the present context. For any irreducible variety X defined over an algebraically closed field k, say a subset of X is *generic* if it contains the complement of a countable union of proper closed subvarieties of X. Note over any uncountable algebraically closed field a generic subset has k-points.

In the statement of **Theorem 1.1** it is assumed that $G = SL_n(k)$ is defined over an uncountable algebraically closed field k. It is evident some condition on k is required, as G is locally finite when k is algebraic over a finite field. However once **Theorem 1.1** has been established, it is possible to use this result to substantially relax the assumptions placed on k. Furthermore, the existence of a single tuple $\omega \in \Omega$ topologically generating G can be used to establish that some larger subset of tuples in Ω generate G topologically.

Theorem 1.2. Let $C_1, \ldots, C_e$ be noncentral conjugacy classes of $G = SL_n(k)$, $n \geqslant 3$, where k is any field. Assume $\sum_{i=1}^{e} \gamma_i \leqslant n(e-1)$, and it is not the case that $e = 2$ and C_1, C_2 are quadratic.

(i) If k is an uncountable algebraically closed field, a generic subset of tuples in Ω generate G topologically.
(ii) If k is any field of characteristic zero, a nonempty open subset of tuples in Ω generate G topologically.

(iii) If k is an algebraically closed field that is not algebraic over a finite field, a dense subset of tuples in Ω generate G topologically.

Similarly if G is any simply connected simple algebraic group defined over a field k, and $\omega \in \Omega$ is a tuple that topologically generates G, natural analogues to conditions $(i) - (iii)$ exist (see **Theorem 6.2** below).

Now let us briefly describe the proof strategy for the main theorem. To begin, pick noncentral conjugacy classes $C_1, \ldots, C_e$ of $G = SL_n(k)$. Let M be a closed subgroup of G, and

$$Y_M := \overline{\{(x_1, \ldots, x_e) \in \Omega \mid \langle x_1, \ldots, x_e \rangle \subset M^g, \text{ for some } g \in G\}}.$$

Up to conjugacy, there exist finitely many maximal positive-dimensional closed subgroups of G (see **Corollary 3** of [46]). Call these subgroups $M_1, \ldots, M_t$. In positive characteristic, a *subfield subgroup* of G is a finite subgroup of the form $G(q)$ with $q = p^a$ where p is the characteristic of the underlying field k. Note each subfield subgroup $G(q)$ is a closed subgroup of G, and every proper closed subgroup of G is contained in either some subfield subgroup, or some positive dimensional maximal subgroup of G (see **Lemma 3.1** of [24]).

From now on we will say that generic tuples in Ω possess a given property if a generic subset of Ω has the property. The overall proof strategy is as follows. Suppose we can show that generic tuples in Ω generate an infinite group, and that generic tuples in Ω are not contained in $\bigcup_{i=1}^{t} Y_{M_i}$. Then generic tuples will generate a subgroup of G contained in no conjugate of a maximal positive-dimensional closed subgroup, or a conjugate of any subfield subgroup. In this case, generic tuples in Ω topologically generate G.

The argument proceeds by induction on the dimension of the natural module. In the base case $n = 3$, the maximal subgroup structure of G is completely understood. Here, for any closed subgroup M of G, **Lemma 2.1** provides the useful bound

$$\dim Y_M \leqslant \dim \prod_{j=1}^{e} (C_j \cap M) + \dim G/M.$$

Then by computing the dimensions of varieties, we show that

$$\dim Y_M \leqslant \dim \prod_{j=1}^{e} (C_j \cap M) + \dim G/M < \dim \Omega$$

for each maximal closed subgroup, and each subfield subgroup M of G. It follows that generic tuples in Ω generate an infinite group not contained in $\bigcup_{i=1}^{t} Y_{M_i}$.

For the inductive argument, slightly more indirect reasoning is required. Using the inductive hypothesis, we infer that some tuple $\omega \in \Omega$ topologically generates a large rank subgroup of G. Making use of this fact, we then show that generic tuples in Ω generate a group having a list of properties that no subfield subgroup nor positive-dimensional maximal subgroup can share. In particular, we show that generic tuples in Ω act irreducibly and primitively on the natural module, and contain "strongly regular" elements of infinite order. (For a given maximal torus T of G, an element of G is said to be *strongly regular* if it has distinct eigenvalues on the root spaces of $Lie(G)$ with respect to T). We then show this implies that generic tuples in Ω topologically generate G.

Finally we turn to applications of the main theorem. The first concerns linear representations of algebraic groups. In what follows let G be a simply connected simple algebraic group defined over an algebraically closed field k of characteristic $p \geqslant 0$. Unless otherwise stated, let V be an irreducible finite dimensional rational kG-module. Recall for any $x \in V$ the point stabilizer of x in G is written $G_x := \{g \in G \mid gx = x\}$. A subgroup H is said to be a *generic stabilizer* of G on V if there exists a nonempty open subset $V_0 \subset V$ such that G_x is a conjugate of H for every $x \in V_0$.

In characteristic zero, it is a well-known result of Richardson [57] that a generic stabilizer exists whenever V is a smooth affine variety (in fact, this result holds when G is reductive). In positive characteristic this fails, although in many cases of interest generic stabilizers do exist, for instance when V is irreducible [25].

The structure of generic stabilizers, and in particular the question of whether a generic stabilizer is finite or trivial, arises frequently in the context of invariant theory. In characteristic zero, the structure theory of generic stabilizers has been carefully investigated. For instance, Èlašvili [14] and Popov [54] classify when a generic stabilizer is finite or trivial. A well-known related result states that a generic stabilizer of G is nontrivial if its algebra of invariant polynomials $k[V]^G$ is free (see **Theorem 8.8** of [56]). For semisimple algebraic groups, the question of when a generic stabilizer is trivial is studied in [14], [15] and [55].

Much recent work concerns the structure of generic stabilizers in arbitrary characteristic. In this context, Guralnick and Lawther [25] recently established that a generic stabilizer is finite if and only if $\dim V > \dim G$.

In addition, for applications to Galois cohomology and essential dimension, it is often necessary to consider a generic stabilizer as a group scheme [3] [19] [35] [50]. Further recent work [17] [18] shows if $\dim V > \dim G$, then in most cases a generic stabilizer is trivial as a group scheme.

As these results relate to the bound given in **Theorem 1.3** below, let us describe this work in slightly greater detail. If G is an algebraic group defined over a field k and $\rho : G \to GL(V)$ is a representation, G is said to act *generically freely* on V if there exists a dense open subset U of V such that for every extension K of k and every $u \in U(K)$, the stabilizer G_u (a closed subgroup-scheme of

$G \times K$) is the group scheme 1. In [17] and [18] it is shown if $\rho : G \to GL(V)$ is a faithful, irreducible representation of a simple algebraic group defined over any algebraically closed field k, then a generic stabilizer acts generically freely on V if and only if $\dim V > \dim G$ and the pair (G, V) does not appear on a short list of exceptional cases (see Tables 4 and 5 in [16]).

For applications, it is often necessary to know not just whether G_v is infinitesimal (i.e. 1 as a group scheme), but whether G_v is trivial as a group scheme. To determine this, in [16] it is established when a generic Lie algebra stabilizer $Lie(G)_v$ is as small as possible (i.e. when $Lie(G)_v = \ker d\rho$).

More precisely, let

$$V^G = \{v \in V | \ gv = v \text{ for all } g \in G\}$$

and for any Lie subalgebra $\mathfrak{s} \subseteq \mathfrak{g}$, let

$$V^{\mathfrak{s}} = \{v \in V | \ d\rho(x)v = 0 \text{ for all } x \in \mathfrak{s}\}.$$

Furthermore let $\rho : G \to GL(V)$ be a representation of G where V has a G-subquotient X with $X^{[\mathfrak{g},\mathfrak{g}]} = 0$. In [16] new bounds $b(G)$ are determined such that if $\dim X > b(G)$, then for generic $v \in V$, $Lie(G)_v = \ker d\rho$.

In **Section 5**, we use **Theorem 1.1** to establish related bounds for algebraic groups, as opposed to their Lie algebras. Let $G = SL_n(k)$ where k is an algebraically closed field of arbitrary characteristic, and let V be a finite dimensional rational kG-module such that $V^G = 0$. (Note in particular we are not assuming that V is irreducible). In this context, we show a generic stabilizer G_v of G is trivial as a group whenever $\dim V > \alpha$. Interestingly this bound α is the same one obtained for the corresponding Lie algebra $\mathfrak{sl}_n(k)$ in [16].

Furthermore, our result relates to a generic stabilizer being trivial as a group scheme. To establish this fact it suffices to show the group stabilizer is generically trivial, and the generic Lie algebra annihilator is zero. (In characteristic zero, it suffices to show the group stabilizer is generically trivial.) Our result is as follows:

Theorem 1.3. Let $G = SL_n(k)$, where $n \geqslant 3$ and k is an algebraically closed field. Assume V is a finite dimensional rational kG-module such that $V^G = 0$. If $\dim V > \frac{9}{4}n^2$, then a generic stabilizer is trivial.

The bound in **Theorem 1.3** is close to best possible. For instance, if $V = \mathrm{Sym}^2(W) \oplus \mathrm{Sym}^2(W)$, where W is the natural module for G, then $\dim V = n(n + 1)$ and a generic stabilizer (in characteristic $p \neq 2$) is a nontrivial elementary abelian group. The generic stabilizer is the generic intersection of the individual generic stabilizers, which are centralizers of graph automorphisms of order 2 inverting a maximal torus of G. It follows the generic stabilizer is a nontrivial elementary abelian 2-group contained in the maximal torus T of G inverted by both of these involutions (see [7], **Lemma 3.6**).

Note in the above example that the module V is reducible. It appears that examples where $\dim V > n^2$ and a generic stabilizer is nontrivial arise only when V is reducible. However when V is irreducible, there are still instances where $\dim V < \frac{9}{4}n^2$ and a generic stabilizer is nontrivial. For example, if $G = SL_9(k)$ and $V = \wedge^3(W)$ where W the natural module for G. Then $\dim V = 84 \leqslant \frac{9}{4} \cdot 9^2$, and a generic stabilizer (in characteristic $p \neq 3$) is a nontrivial elementary abelian group. See Table 1 of [25] for this and other examples.

A second application of our main theorem concerns a classical group-theoretic problem: the random generation of finite simple groups. The following two questions date back to the nineteenth century:

(i)   does the probability that two random elements of Alt_n generate Alt_n tend to 1 as $n \to \infty$?

(ii)   which finite simple groups appear as quotients of the modular group $PSL_2(\mathbb{Z})$?

The first question was posed by Netto in 1892 and answered affirmatively by Dixon [12] in 1969. In the same paper, Dixon conjectures that for *any* finite simple group H, the probability that two random elements of H generate H tends to 1 as the size of H tends to infinity. After some effort, this conjecture was proven by Kantor-Lubotzky [32] and Liebeck-Shalev [47] in the 1990s using probabilistic methods.

The positive solution of Dixon's conjecture led to many more fine-grained questions concerning the random generation of finite simple groups. A natural refinement of the above question concerns whether two random elements of prime orders r and s generate H as $|H| \to \infty$. The property that two elements of orders r and s generate H with probability tending to 1 as $|H|$ tends to infinity is called *random (r, s)-generation*. Perhaps surprisingly given the strength of the statement, this property has also been found to hold in many cases [48], although counterexamples exist [49], and some questions remain. In particular, by work of Liebeck and Shalev [48] it is known that finite simple classical groups have random (r, s)-generation when $(r, s) \neq (2, 2)$ and the rank of the group is large enough (depending on the primes r and s).

The weaker question of determining whether *any* elements of orders r and s, and in particular whether any elements of orders 2 and 3, generate a finite simple group is also a classical group-theoretic problem. This property is called (r, s)-*generation*. Its geometric background and long history in the literature are discussed in [49].

Although it is not immediately obvious, question (ii) listed above concerns the $(2, 3)$-generation of finite simple groups. As $PSL_2(\mathbb{Z})$ is isomorphic to the free product $C_2 * C_3$, the question may be rephrased in the following form: which finite simple groups are $(2, 3)$-generated?

Much progress on this problem has been made in recent years. In [52], it is shown all finite simple exceptional groups (except Suzuki groups) are $(2, 3)$-generated. Suzuki groups are obvious exceptions because they do not contain any elements of order 3. For the classical groups, recent advances have been made by proving stronger random $(2, 3)$-generation results. In [49] it is shown that all finite simple classical groups not of the form $PSp_4(q)$ have random $(2, 3)$-generation. Consequently, aside from $PSp_4(q)$, all but finitely many finite simple classical groups appear as quotients of $PSL_2(\mathbb{Z})$.

For groups of type $PSp_4(q)$, the random $(2, 3)$-generation results are more nuanced. First, groups of the form $PSp_4(p^f)$ with $p \in \{2, 3\}$ are not $(2, 3)$-generated for any f. For $PSp_4(p^f)$, with $p \geqslant 5$, the probability that two random elements of orders 2 and 3 generate $PSp_4(p^f)$ tends to $\frac{1}{2}$ as the size of the group goes to infinity [49]. In recent work [27] it is shown that all exceptional groups of Lie type (except Suzuki groups) also have random $(2, 3)$-generation.

Using random $(2, 3)$-generation as a guide, one might expect (modulo a few small rank exceptions) that random (r, s)-generation holds for all finite simple groups, with r, s independent of the rank of the group. In **Section 5**, we use topological generation to prove this fact for linear and unitary groups, hence strengthening results in [48]. In work with collaborators [5] [6], this property is established for the remaining finite simple groups.

In what follows, the rank of a finite unitary group $G(q)$ is defined to be its untwisted rank, which coincides with the rank of the ambient simple algebraic group G. Furthermore, the characteristic of $\mathbb{F}_q$ is assumed to be fixed.

Theorem 1.4. Fix a positive integer n, and let r, s be primes with $(r, s) \neq (2, 2)$. Let $G(q)$ be a finite simple linear or unitary group of fixed rank n over the finite field $\mathbb{F}_q$ and assume $G(q)$ contains elements of orders r and s. Let x and y be random elements of orders r and s in $G(q)$. Then the probability that x and y generate $G(q)$ tends to 1 as $q \to \infty$.

2 Preliminary results

Before proving **Theorem 1.1**, it will be helpful to establish some general facts about algebraic groups required in our arguments. This background material will be stated in slightly greater generality than necessary, so that it may be applied in future work.

In this section (unless otherwise stated) let G be a simply connected simple algebraic group defined over an algebraically closed field of characteristic $p \geqslant 0$. When G is a classical group, let V be the natural module for G. Finally, let $C_1, \ldots, C_e$ be noncentral conjugacy classes of G, and $\Omega := C_1 \times \ldots \times C_e$. Say that a tuple $(x_1, \ldots, x_e) \in G^e$ has a given group property if the closure of $\langle x_1, \ldots, x_e \rangle$ has that property.

For any closed subgroup M of G, let:

- $\Delta := \prod_{j=1}^{e} (C_j \cap M)$
- $X := \bigcup_{g \in G} (M^g \times \ldots \times M^g) \subset G^e$
- $\varphi : G \times \Delta \to X$ be the map $(g, (x_1, \ldots, x_e)) \mapsto (x_1^g, \ldots, x_e^g)$
- $Y_M := \{(x_1, \ldots, x_e) \in \Omega \mid \langle x_1, \ldots, x_e \rangle \subset M^g, \text{ for some } g \in G\}$

Note for any closed subgroup M of G, $(x_1^g, \ldots, x_e^g) \in im(\varphi)$ if and only if $(x_1, \ldots, x_e) \in \Omega$ and $\langle x_1, \ldots, x_e \rangle \subset M^g$ for some $g \in G$. In particular, $im(\varphi) = Y_M$.

There exist finitely many conjugacy classes of maximal positive-dimensional closed subgroups of G (see **Corollary 3** in [46]). Call these subgroups $M_1, \ldots, M_t$. As discussed in **Section 1**, the overall strategy is to prove that generic tuples in Ω topologically generate an infinite group and that generic tuples in Ω are not contained in $\bigcup_{i=1}^{t} Y_{M_i}$. The next result reduces proving the base case $n = 3$ of the main theorem to performing calculations on the dimensions of explicit varieties.

Lemma 2.1. Let M be a closed subgroup of G. We have

$$\dim Y_M \leqslant \dim \Delta + \dim G/M.$$

Proof. To begin, note any $(x_1, \ldots, x_e)^g \in im(\varphi)$ will be in the same orbit as $(x_1, \ldots, x_e) \in \Delta$, and hence their fibers will have the same dimension. So without loss of generality, assume $(x_1, \ldots, x_e) \in \Delta$ and consider the map $\varphi' : M \times \Delta \to \Delta$, $(g, (x_1, \ldots, x_e)) \mapsto (x_1^g, \ldots, x_e^g)$. Clearly $im(\varphi') = \Delta$, and hence every fiber of φ has dimension at least $\dim M$. By the fiber theorem (see **Corollary 4** of EGA III [23]), $\dim Y_M \leqslant \dim \Delta + \dim G/M$. $\square$

Remark 2.2. Over an uncountable algebraically closed field, the above lemma reduces proving generic tuples in Ω topologically generate G to establishing that $\dim \Omega > \dim \Delta + \dim G/M$, for each maximal positive-dimensional closed subgroup M of G.

The following is a version of Burnside's Lemma for algebraic groups, which allows us to recast **Lemma 2.1** in a more geometric light.

Lemma 2.3. Consider the action of G on the coset variety G/M. Then for $x \in M$,

$$\dim G/M - \dim (G/M)^x = \dim C - \dim (C \cap M)$$

where $(G/M)^x$ is the set of fixed points of x, and C is the conjugacy class of x in G.

Proof. This is **Proposition 1.14** in [44]. □

This immediately yields the following useful bound on fixed point spaces.

Lemma 2.4. Write $C_i = x_i^G$ for $i = 1, \ldots, e$. If $\Omega \subset Y_M$, then

$$\sum_{j=1}^{e} \dim (G/M)^{x_j} \geqslant (e-1) \cdot \dim G/M.$$

Proof. Assume $\Omega \subset Y_M$. By **Lemma 2.1**, we have

$$\dim \Omega \leqslant \sum_{j=1}^{e} \dim (C_j \cap M) + \dim G/M.$$

Summing from 1 to e, **Lemma 2.3** yields

$$\dim \Omega = \sum_{j=1}^{e} \dim G/M - \sum_{j=1}^{e} \dim (G/M)^{x_j} + \sum_{j=1}^{e} \dim (C_j \cap M),$$

and by combining these facts we deduce that

$$\sum_{j=1}^{e} \dim (G/M)^{x_j} \geqslant (\sum_{j=1}^{e} \dim G/M) - \dim G/M.$$

□

For the inductive argument required in the proof of **Theorem 1.1**, we will use the inductive hypothesis to infer the existence of a tuple $\omega \in \Omega$ which topologically generates a large rank subgroup of $SL_n(k)$. We will then use this fact to show that generic tuples in Ω generate a group having a list of properties that no subfield subgroup nor positive-dimensional maximal subgroup can share.

Typically these arguments proceed in two stages. The first is to prove the existence of a tuple $\omega \in \Omega$ having some desirable property, and the second is to show that this property is either an open or generic condition.

We will make use of several facts from the literature. Let V be a rational kG-module. Furthermore let $\mathcal{R}_{d,e}(V)$ be the set of tuples in G^e fixing a given d-dimensional subspace of V, $\mathcal{I}_{d,e}(V)$ be the set of tuples in G^e that fix no d-dimensional subspace of V, and $\mathcal{I}_e(V)$ be the set of tuples in G^e that act irreducibly on V. The following lemma is well-known, and is used to show an open subset of tuples in Ω generate a group acting irreducibly on the natural module.

Lemma 2.5.   Let V be a rational kG-module. Then

(i) $\mathcal{R}_{d,e}(V)$ is a closed subvariety of G^e.
(ii) $\mathcal{I}_{d,e}(V)$ is an open subvariety of G^e.
(iii) $\mathcal{I}_e(V)$ is an open subvariety of G^e.

  Proof.   Parts *(i)* and *(ii)* follow from **Lemma 11.1** in [29]. For part *(iii)*, assume $\dim V = n$. Then $\cup_{1 \leqslant d \leqslant n} \mathcal{I}_{d,e}(V) = \mathcal{I}_e(V)$.   □

An extension of **Lemma 2.5** proved in [2] will also be useful for establishing properties of topological generation. Let q be a power of a prime p. Recall for a given Frobenius endomorphism F_q of G, that the fixed point group $G^{F_q} = G(q)$ is a finite group of Lie type over $\mathbb{F}_q$.

Lemma 2.6.   There exists a finite collection $V_1, \ldots, V_t$ of irreducible finite-dimensional kG-modules with the property that if a proper closed subgroup of G acts irreducibly on V_i for each i, then it is finite and conjugate to $G(q)$ for some q. In particular, if $char(k) = 0$, there are no such subgroups.

  Proof.   This is **Lemma 4.2** in [2].   □

Remark 2.7.   By **Lemma 2.5** *(iii)* acting irreducibly on a single (and hence any finite collection) of rational kG-modules is an open condition, so **Lemma 2.6** shows the set of tuples in G^e topologically generating a group containing a conjugate of some fixed subfield subgroup $G(q_0)$ of G is open in G^e. In $char(k) = 0$, **Lemmas 2.5** and **2.6** show that topologically generating G is an open condition.

To establish the remaining generic properties of Ω it will be necessary to move between properties holding generically on a simple algebraic group G and properties holding generically on G^e. In what follows, let F_e refer to the free group of rank e.

Lemma 2.8. For any nontrivial word $w \in F_e$, the word map $\varphi_w : G^e \to G$ which sends $(x_1, \ldots, x_e) \mapsto w(x_1, \ldots, x_e)$ is dominant. In particular, $Z \subseteq G$ is a generic subset of G if any only if

$$\{(x_1, \ldots, x_e) \in G^e \mid w(x_1, \ldots, x_e) \in Z\}$$

is a generic subset of G^e.

> **Proof.** This is a result of Borel, and appears as **Proposition 2.5** in [2]. □

Let us say a tuple $(x_1, \ldots, x_e) \in G^e$ possesses a property of an element of G (has infinite order, is regular semisimple, etc.) if there is a nontrivial word $w \in F_e$ such that $w(x_1, \ldots, x_e)$ has that property.

Perhaps surprisingly, some effort is required to show that generic tuples in Ω generate an infinite group (and therefore are not contained in a subfield subgroup of G). The reason for this difficulty is that the property of having infinite order, or of generating an infinite subgroup, is not an open condition. However if we assume G is defined over an uncountable algebraically closed field, then having infinite order is a nonempty generic condition.

Lemma 2.9. Let k be an uncountable algebraically closed field. Then generic tuples in G^e have infinite order. Furthermore if some $\omega \in \Omega$ has infinite order, generic tuples in Ω share this property.

> **Proof.** The set $\Theta_n = \{x \in G \mid x^n = 1\}$ of elements of order dividing n is closed in G. Over an uncountable field, $(\bigcup_{n \geqslant 1} \Theta_n)^c$ is a generic subset of G. Applying **Lemma 2.8**, generic tuples in G^e have infinite order. If some $\omega \in \Omega$ has infinite order, then $\Omega \cap \Theta_n$ is a proper closed subvariety of Ω for each n. □

Furthermore over an uncountable algebraically closed field, the existence of a single tuple $\omega \in \Omega$ topologically generating G implies a generic subset of Ω possess the same property.

Lemma 2.10. Let G be defined over an uncountable algebraically closed field k. If there is a tuple $\omega \in \Omega$ topologically generating G, then generic tuples in Ω generate G topologically.

> **Proof.** First assume G has characteristic $p = 0$. Then **Lemmas 2.5** and **2.6** and **Remark 2.7** show that the existence of a single tuple $\omega \in \Omega$ topologically generating G implies an open dense subset of tuples in Ω share this property (this is also true over any algebraically closed field).

In positive characteristic, **Lemma 2.6** shows there is a proper closed subvariety of G^e outside of which a tuple $\omega \in G^e$ generates a dense subgroup of G if and only if it generates an infinite subgroup of G. By **Lemma 2.9**, the existence of a single tuple $\omega \in \Omega$ generating an infinite subgroup implies that generic tuples in Ω share this property. □

Now recall G acts primitively on a kG-module V if it acts irreducibly and preserves no non-trivial direct sum decomposition of V. The following lemmas establish that the set of tuples in G^e generating a primitive subgroup is an open subset of G^e.

Lemma 2.11. Let G be a subgroup of $GL(V)$, with V an n-dimensional vector space. Then G acts primitively on V if and only if

(i) $\quad$ G acts irreducibly on V, and
(ii) $\quad$ For each proper divisor d of n, no subgroup H with $[G : H] = n/d$ fixes a d-dimensional subspace of V.

> **Proof.** To start we may assume that G acts irreducibly on V or else there is nothing to prove. If G acts imprimitively and irreducibly, then $V = V_1 \oplus \ldots \oplus V_{n/d}$ with $n/d > 1$, and G permutes the V_i transitively. If H is the stabilizer of V_1, then $[G : H] = n/d$.
>
> Conversely, suppose there is a subgroup H of G such that $[G : H] = n/d$, $d < n$, and H fixes a d-dimensional subspace V_1 of V. For $1 \leq i \leq n/d$, let V_i be the conjugates of V_1 under G. Since G acts irreducibly, $V = V_1 \oplus \ldots \oplus V_{n/d}$, and G is imprimitive. □

Next let n be any positive integer and fix a divisor d of n. Since the free group F_e is finitely generated, it has only finitely many subgroups of index d. Label these subgroups $H_1(d), \ldots, H_t(d)$. For each such $H_i(d)$, $1 \leq i \leq t$, choose a collection of generators $w_{i1}, \ldots, w_{is}$ of $H_i(d)$.

Lemma 2.12. Let G be a simple subgroup of $GL(V)$, with V an n-dimensional vector space defined over an algebraically closed field. The set of tuples $\omega \in \Omega$ acting primitively on V forms a Zariski open subset of G^e.

> **Proof.** To begin, note the set of tuples $(x_1, \ldots, x_e) \in G^e$ such that $\langle x_1, \ldots, x_e \rangle$ fixes a common d-dimensional subspace of V is the same as the set of $(x_1, \ldots, x_e) \in G^e$ such that $\overline{\langle x_1, \ldots, x_e \rangle}$ fixes a common d-dimensional subspace of V. By **Lemma 2.5**, the set of tuples in G^e topologically generating a subgroup acting reducibly on V is closed.

Now suppose $(g_1, \ldots, g_e) \in G^e$ topologically generates a subgroup J of G that acts irreducibly and imprimitively on V. Then by **Lemma 2.11**, for some proper divisor d of n there is a subgroup $J_1 \subset J$ such that $[J : J_1] = n/d$, and J_1 fixes a d-dimensional subspace of V. Choose a surjection $\varphi : F_e \to J$ such that $\varphi^{-1}(J_1) = H_i(d)$ for some $1 \leqslant i \leqslant t$. Then $\langle w_{i1}(g_1, \ldots, g_e), \ldots, w_{is}(g_1, \ldots, g_e) \rangle$ fixes a common d-dimensional subspace of V.

By **Lemma 2.5**, the set of tuples $(x_1, \ldots, x_e) \in G^e$ such that the closure of $\langle w_{i1}(x_1, \ldots, x_e), \ldots, w_{is}(x_1, \ldots, x_e) \rangle$ fixes a common d-dimensional subspace of V is closed in G^e. Then considering each subgroup $H_1(d), \ldots, H_t(d)$ for each divisor d of n, we find the set of tuples $(x_1, \ldots, x_e) \in G^e$ topologically generating an imprimitive subgroup is closed. $\square$

Remark 2.13. When G is a simple subgroup of $GL(V)$, $e > 1$, and k is not algebraic over a finite field, there are tuples $(x_1, \ldots, x_e) \in G^e$ topologically generating G (see **Corollary 3.4** in [24]). So the open subset described above is nonempty. Of course this set is empty when $e = 1$.

To conclude this section, let us focus on conditions for topological generation for specific classical groups. These results will be used in the proof of **Theorem 1.1**, and later in [6]. In what follows, assume G is a simply connected simple algebraic group of classical type, and V is the natural module for G. Recall for a conjugacy class C_j of G that γ_j is the dimension of the largest eigenspace on V of a representative of C_j, and $\Omega = C_1 \times \cdots \times C_e$. Set $n = \dim V$. The following lemma gives a necessary condition for topological generation.

Lemma 2.14. If $\sum_{i=1}^{e} \gamma_i > n(e - 1)$, then no tuple $\omega \in \Omega$ topologically generates G.

Proof. For any element $x \in C_i$, let E_{α_i} be a γ_i-dimensional eigenspace of x on V. We claim

$$\dim\left(\bigcap_{i=1}^{e} E_{\alpha_i}\right) = \sum_{i=1}^{e} \gamma_i - n(e - 1).$$

This is easily checked by induction on the number of classes e. If $e = 2$ and $\gamma_1 + \gamma_2 > n$, then clearly $\dim(E_{\alpha_1} \cap E_{\alpha_2}) = \gamma_1 + \gamma_2 - n$. So assume $\sum_{i=1}^{k} \gamma_i > n(k - 1)$, and

$$\dim\left(\bigcap_{i=1}^{k} E_{\alpha_i}\right) = \sum_{i=1}^{k} \gamma_i - n(k - 1).$$

Then

$$\dim \left(\bigcap_{i=1}^{k} E_{\alpha_i} \cap E_{\alpha_{k+1}} \right) = \left(\sum_{i=1}^{k} \gamma_i - n(k-1) \right) + \gamma_{k+1} - n.$$

Letting $e = k + 1$, it follows that

$$\dim \left(\bigcap_{i=1}^{e} E_{\alpha_i} \right) > 0$$

whenever $\sum_{i=1}^{e} \gamma_i > n(e-1)$. In this case, every tuple $\omega \in \Omega$ topologically generates a reducible subgroup on the natural module, and hence no tuple in Ω will generate G topologically. □

Next assume M is the stabilizer of a 1-dimensional totally singular subspace of V (so M is a maximal parabolic subgroup of G). For linear and symplectic groups, all 1-dimensional subspaces of V are totally singular. Furthermore in these cases G acts transitively on the relevant set of 1-spaces, so G/M is isomorphic to projective space $P_1(V)$ as a variety. Hence to show there exists some tuple $\omega \in \Omega$ fixing no 1-dimensional subspace of V, it suffices to show

$$\Omega \not\subset X := \bigcup_{g \in G} (M^g \times \ldots \times M^g) \subset G^e.$$

(For orthogonal groups, 1-dimensional non-singular subspaces must also be considered.)

Lemma 2.15. Let $G = SL_n(k)$ or $Sp_n(k)$. Then $\sum_{j=1}^{e} \gamma_j \leqslant n(e-1)$ if and only if there is a tuple $\omega \in \Omega$ fixing no 1-dimensional subspace of V.

Proof. To begin assume $\sum_{j=1}^{e} \gamma_j > n(e-1)$. For each conjugacy class C_i, pick some $x_i \in C_i$ and let E_{α_i} be a maximal dimensional eigenspace of x_i on V. By the argument given in **Lemma 2.14**, it follows that $\dim \left(\bigcap_{i=1}^{e} E_{\alpha_i} \right) > 0$. In particular, there exists a 1-dimensional subspace E of V such that $E \subseteq \bigcap_{i=1}^{e} E_{\alpha_i}$, and $\omega = (x_1, \ldots, x_e)$ fixes E. Hence every tuple $\omega \in \Omega$ fixes some 1-dimensional subspace of V.

For the converse direction, assume $\Omega \subset X$ where M is the stabilizer of a 1-dimensional subspace of V. Let $\varphi : G^e \times G/M \to G^e$ be the natural projection map. Since M is parabolic, G/M is a complete variety and φ is a closed map. However, $\varphi(Z) = X$ where

$$Z = \{(x_1, \ldots, x_e, v) \mid x_1 v = \ldots = x_e v = v\}.$$

So X is closed, and in particular

$$\Omega \subset Y_M = \overline{\{(x_1, \ldots, x_e) \in \Omega \mid \langle x_1, \ldots, x_e \rangle \subset M^g, \text{ for some } g \in G\}}.$$

Applying **Lemma 2.3**,

$$\sum_{j=1}^{e} \dim\, (G/M)^{x_j} \geqslant (e-1)\cdot \dim\, G/M.$$

Note the irreducible components of $(G/M)^x$ are projective spaces associated with each eigenspace of x. In particular the dimension of $(G/M)^x$ is one less than the dimension of the largest eigenspace of x on V.

So $\dim\, (G/M)^{x_j} = \gamma_j - 1$, and $\dim G/M = n - 1$. Applying this information to the above inequality yields

$$(\sum_{j=1}^{e} \gamma_j) - e \geqslant (e-1)(n-1).$$

This implies that $\sum_{j=1}^{e} \gamma_j > n(e-1)$.   □

Remark 2.16.   Since Ω is an irreducible variety, **Lemmas 2.5** and **2.15** imply that under the conditions given in **Theorem 1.1**, an open subset of tuples $\Omega_0 \subset \Omega$ fixes no 1-dimensional subspace of V.

Remark 2.17.   No restrictions are placed on the conjugacy classes $C_1, \ldots, C_e$ of G since only 1-dimensional subspaces of V are being considered. When moving to the stabilizers of higher dimensional subspaces, different arguments are required to compute $\dim\, (G/M)^x$ depending on whether the $x \in C$ are semisimple, unipotent, or mixed. This issue arises when providing inductive arguments to establish topological generation for symplectic and orthogonal groups.

Now pick a maximal torus T of G and let $Ad : G \rightarrow GL(Lie(G))$ be the adjoint representation. Note $Lie(G)$ embeds in $V \otimes V^*$, and the rank of G is the dimension of T. An element $x \in G$ is *regular* if the dimension of its centralizer is equal to the rank of the group, and *regular semisimple* if it is both regular and diagonalizable. Call a regular semisimple element of G *strongly regular* if it has distinct eigenvalues on the root spaces of $Lie(G)$ with respect to T. In other words, $t \in T$ is strongly regular if for any two distinct roots α_i, α_j of G, $\alpha_i(t) \neq \alpha_j(t)$. Note if an element $t \in T$ has eigenvalues $\lambda_1, \ldots, \lambda_n$ on V, then t has eigenvalues $\lambda_i \cdot \lambda_j^{-1}$ on $V \otimes V^*$, for $1 \leqslant i \leqslant n, 1 \leqslant j \leqslant n$. Hence if for every $1 \leqslant i, j, k, l \leqslant n$, we have that $\lambda_i \cdot \lambda_j^{-1} = \lambda_k \cdot \lambda_l^{-1}$ implies $i = j$ and $k = l$, then t is strongly regular.

The set of strongly regular elements is clearly open in G (since the complement is defined by finitely many polynomial equations) and nonempty. Recall a tuple $(x_1, \ldots, x_e) \in \Omega$ is said to be strongly regular if there is a nontrivial word $w \in F_e$ such that $w(x_1, \ldots, x_e)$ has that property. Applying **Lemmas 2.8** and **2.9**, being strongly regular and having infinite order are generic properties of G^e (defined over an uncountable algebraically closed field). Hence if some tuple $\omega \in \Omega$ is strongly regular of infinite order, then generic tuples in Ω share this property.

Proposition 2.19 below provides conditions sufficient to ensure the existence of a tuple $\omega \in \Omega$ topologically generating $G = SL_n(k)$. In the next section it is shown that these conditions may in fact be satisfied. We first recall a useful fact from the literature. Here, a maximal rank subgroup is a subgroup that contains a maximal torus of G.

Lemma 2.18. Let $G = SL_n(k)$, $n \geqslant 3$, and let H be a proper maximal rank subgroup of G. Then the connected component H° acts irreducibly on the natural module for G.

> Proof. This follows from **Section 6** of [51], and the main theorem of [45]. □

Proposition 2.19. Let $C_1, \ldots, C_e$ be noncentral conjugacy classes of $G = SL_n(k)$, where k is an uncountable algebraically closed field, and $n \geqslant 3$. Assume

(i) there is a tuple $\omega \in \Omega$ that acts irreducibly and primitively on the natural module V, and

(ii) there is a tuple $\omega \in \Omega$ that is strongly regular of infinite order.

Then there is a tuple $\omega \in \Omega$ topologically generating G.

> Proof. Assume that conditions *(i)* and *(ii)* both hold. Then by **Lemmas 2.5** and **2.11**, a non-empty open subset of tuples $\Omega_0 \subset \Omega$ share property *(i)*. As being strongly regular of infinite order is a generic property, a generic subset of tuples in Ω will satisfy condition *(ii)*. In particular, some $\omega \in \Omega$ satisfies both conditions. Let H be the closure of the subgroup generated by this tuple. Note $Lie(H)$ is a sum of eigenspaces for a strongly regular element $x \in H$, and so is a direct sum $T_0 \oplus N_{\alpha_1} \oplus \ldots \oplus N_{\alpha_j}$ of some collection of nontrivial weight spaces for T. Furthermore, $T_0 \subseteq T$, where $T = C_G(x)$.
>
> Assume $Lie(H) \neq Lie(G)$. Then $Lie(H)$ is invariant under T, and the closure of $\langle H, T \rangle$ is a proper maximal rank subgroup of G. However by **Lemma 2.18** this is impossible. Hence $Lie(H) = Lie(G)$ which in turn implies $H = G$. So there exists a tuple $\omega \in \Omega$ topologically generating G. □

We are now in a position to prove the main theorem. In this section, let $C_1, \ldots, C_e$ be noncentral conjugacy classes of $G = SL_n(k)$, where k is an uncountable algebraically closed field of characteristic $p \geqslant 0$, and V is the natural module for G. Pick a class C from $C_1, \ldots, C_e$, and $x \in C$. Assume x has eigenvalues $\lambda_1, \ldots, \lambda_t$ on V, and recall x is conjugate to a block diagonal matrix of the form

$$J = \begin{bmatrix} J_{r_1}(\lambda_1) & & \\ & J_{r_2}(\lambda_2) & \\ & & J_{r_t}(\lambda_t) \end{bmatrix}$$

where

$$J_{r_i}(\lambda_i) = \begin{bmatrix} \lambda_i & 1 & 0 & \ldots & 0 \\ 0 & \lambda_i & 1 & \ldots & 0 \\ \cdots & & & & \\ \cdots & & & & \\ 0 & \ldots & 0 & \lambda_i & 1 \\ 0 & \ldots & 0 & 0 & \lambda_i \end{bmatrix}.$$

We say that $J_{r_i}(\lambda_i)$ is a *Jordan block* of x, where r_i indicates the size of the Jordan block, and λ_i is the corresponding eigenvalue. Furthermore we say $J_{r_1}(\lambda_1) \oplus \ldots \oplus J_{r_t}(\lambda_t)$ is the *Jordan form* of x. Note in this definition we are not assuming that the λ_i are distinct.

Finally, recall a conjugacy class C of G is *quadratic* if a representative $x \in C$ has a degree two minimal polynomial. Note elements in a quadratic class C have Jordan form $J_{r_1}(\lambda_1) \oplus \ldots \oplus J_{r_t}(\lambda_t)$, where $r_i \leqslant 2$ for $1 \leqslant i \leqslant t$, and x has at most two distinct eigenvalues on V.

The "only if" direction of **Theorem 1.1** follows from **Lemma 2.14**, in combination with the following well-known fact from linear algebra.

Lemma 3.1.   Suppose $e = 2$ and C_1, C_2 are quadratic conjugacy classes of G. If $n \geqslant 3$, then every tuple $\omega \in \Omega$ generates a reducible subgroup of V.

       Proof.   This follows from **Lemma 3.14** in [7].   □

For the converse direction, the argument proceeds by induction on the dimension of the natural module. For both the base case and the induction, it is helpful to record some information about the maximal subgroup structure of $SL_n(k)$. This follows from more general work on the maximal subgroups of classical groups carried out by Aschbacher [1], Liebeck and Seitz [45], and others.

In outline, the results state that the positive-dimensional maximal closed subgroups of classical algebraic groups arise in a collection of geometric families

$$C = C_1 \cup C_2 \cup C_3 \cup C_4 \cup C_6$$

and a collection S of almost simple irreducibly embedded subgroups. The geometric families can be described in a uniform fashion in terms of their action on the natural module V. Roughly speaking, the collection C_1 comprises subgroups that stabilize a proper subspace of V, the members of C_2 stabilize an orthogonal decomposition of V, members of C_3 stabilize a totally singular decomposition of V, members of C_4 stabilize a tensor decomposition of V, and members in C_6 are normalizers of classical groups. The remaining positive dimensional maximal subgroups S are less easily described in a systematic fashion, but have the nice property that their connected components are simple groups modulo scalars. See [45] for further details on these subgroup collections.

Note that subfield subgroups are finite closed subgroups acting irreducibly and tensor indecomposably on the natural module. To exclude the possibility that a tuple $\omega \in \Omega$ topologically generates a group contained in a subfield subgroup, it clearly suffices to show that the tuple topologically generates an infinite subgroup.

The following table lists the approximate structures of the maximal geometric subgroups of $G = SL_n(k)$. Note the collection C_3 described above is empty for $SL_n(k)$, and P_m refers to a parabolic subgroup of $SL_n(k)$, which stabilizes an m-dimensional subspace of V.

Table 1.   The maximal geometric subgroups of $SL_n(k)$

Class	structure	conditions
C_1	P_m	$1 \leqslant m \leqslant n - 1$
C_2	$GL_m \wr S_t$	$n = mt, t \geqslant 2$
C_4	$GL_{n_1} \otimes GL_{n_2}$	$n = n_1 n_2,$
		$2 \leqslant n_1 < n_2$
	$(\otimes_{i=1}^{t} GL_m).S_t$	$n = m^t,$
		$m \geqslant 3, t \geqslant 2$
C_6	Sp_n	n even
	SO_n	$p \neq 2$

Theorem 3.2. For $G = SL_n(k)$, $n \geqslant 2$, the conjugacy classes of maximal closed geometric subgroups of G are listed in Table 1.

Proof.   See the main theorem of [45].   □

Corollary 3.3. Let M be a maximal positive dimensional closed subgroup of $SL_3(k)$. Then one of the following holds:

(i) M is irreducible and primitive. In this case $M° \cong SO_3(k)$.
(ii) M is irreducible and imprimitive. In this case $M \cong (GL_1(k) \wr S_3) \cap SL_3(k)$ is the normalizer of a maximal torus.
(iii) M is reducible. In this case M is the stabilizer of a line or a hyperplane on the natural module.

> **Proof.**   This follows from **Theorem 3.2**, and the fact S is empty for $SL_3(k)$.   □

For the base case of our argument, it will be helpful to record the dimensions of noncentral conjugacy classes in $SL_3(k)$.

Lemma 3.4.   Let C be a noncentral conjugacy class of $SL_3(k)$. If C is quadratic then $\dim C = 4$. Otherwise, $\dim C = 6$.

> **Proof.**   Regular elements in $SL_3(k)$ have two dimensional centralizers, and hence if C is a conjugacy class of regular elements, $\dim C = 6$. If C is not a conjugacy class of regular elements, then C is quadratic and representatives of C either have Jordan form $x = J_2(\lambda_1) \oplus J_1(\lambda_2)$, or $x = J_1(\lambda_1) \oplus J_1(\lambda_1) \oplus J_1(\lambda_2)$. In either case, $\dim C_G(x) = 4$ and $\dim C = 8 - 4 = 4$.   □

The following theorem establishes the base case $n = 3$ of **Theorem 1.1**. For the case $n = 2$, see Theorem 4.5 below.

Theorem 3.5.   Let $C_1, \ldots, C_e$ be noncentral conjugacy classes of $G = SL_3(k)$, where k is an uncountable algebraically closed field. Assume $\sum_{j=1}^{e} \gamma_j \leqslant 3(e - 1)$, and it is not the case that $e = 2$ and C_1, C_2 are quadratic. Then there is a tuple $\omega \in \Omega$ topologically generating G.

> **Proof.**   First we show generic tuples in Ω generate an infinite subgroup of G. In view of **Lemma 3.4**, we note that $\dim \Omega \geqslant 10$.
>
> Assume $p > 0$, and M is a subfield subgroup of G. Pick a class C from the list $C_1, \ldots, C_e$. Clearly $\dim (C \cap M) = 0$ and $\dim G/M = 8$. So $\dim \Omega \geqslant 10 > \dim Y_M$, and hence Y_M is a proper closed subvariety of Ω. Now let $M_1, M_2, \ldots$ be an enumeration of all subfield subgroups of G up to conjugacy. Then $\bigcup_{i \geqslant 1} Y_{M_i}$ is a countable union of proper closed subvarieties of Ω. So generic tuples in Ω are contained in $(\bigcup_{i \geqslant 1} Y_{M_i})^c$, and hence generate an infinite subgroup.

Recall that $\Delta = \prod_{j=1}^e (C_j \cap M)$. Working in arbitrary characteristic, it now only remains to show that

$$\dim \Omega > \dim \Delta + \dim G/M$$

for each maximal positive-dimensional closed subgroup M of G. Once this is established, it follows that generic tuples in Ω generate an infinite group contained in no positive-dimensional maximal closed subgroup, and hence generic tuples in Ω generate G topologically. Applying **Corollary 3.3**, it suffices to treat the following cases.

(i) $M = SO_3(k)$. Pick a class C from $C_1, \ldots, C_e$ such that $M \cap C$ is nonempty. Then $M \cap C$ is a finite union of M-classes, each of which has dimension 2, so $\dim (M \cap C) = 2$. As $\dim G/M = 5$, it follows $\dim \Omega > 2e + 5$.

(ii) $M = N_G(T)$ is the normalizer of a maximal torus T. Pick a class C from $C_1, \ldots, C_e$. Note $M^\circ = T$, so $\dim G/M = 6$ and $\dim (M \cap C) \leqslant 2$. Applying **Lemma 3.4**, $\dim \Omega > 2e + 6$ unless (a) $e = 2$ and at most one of C_1, C_2 are non-quadratic, or (b) $e = 3$ and each C_i is quadratic.

        If C is quadratic, then either $M \cap C$ is finite or C is a class of involutions. If $M \cap C$ is finite the desired inequality immediately holds. So assume C is a class of involutions where $M \cap C$ is positive-dimensional. Then $M \cap C$ is a union of classes of outer involutions in M (there are two such classes if characteristic $p \neq 2$, and one when $p = 2$). These outer involutions have a one dimensional centralizer in M. Hence if C is quadratic, $\dim (M \cap C) \leqslant 1$. It follows that $\dim \Omega > \dim \Delta + \dim G/M$.

(iii) M is reducible. Applying **Lemmas 2.5** and **2.15**, we see that generic tuples in Ω do not fix a 1-space or hyperplane (by considering the dual space). $\qquad\qquad\qquad\qquad\qquad\qquad$ □

Next we turn to the inductive argument. In order to make use of the inductive hypothesis, we will need to relate the conjugacy classes $C_1, \ldots, C_e$ of $SL_n(k)$ to classes $C_1', \ldots, C_e'$ of $SL_{n-1}(k)$. It will also be necessary to show the inequality $\sum_{j=1}^e \gamma_j' \leqslant (n-1)(e-1)$ holds, where γ_j' is the dimension of the largest eigenspace of a representative of C_j' on the natural module for $SL_{n-1}(k)$.

To carry out this process, pick a class C in $C_1, \ldots, C_e$, and some element $x \in C$. Assume x has Jordan form $J_{r_1}(\lambda_1) \oplus \ldots \oplus J_{r_t}(\lambda_t)$ on the natural module, and λ_j is an eigenvalue corresponding to a maximal dimensional eigenspace E_{λ_j}. Finally, let $J_{r_j}(\lambda_j)$ be a Jordan block such that $r_j \leqslant r_i$ for all Jordan blocks $J_{r_i}(\lambda_i)$ with $\lambda_i = \lambda_j$.

Now replace the Jordan block $J_{r_j}(\lambda_j)$ in the Jordan form of x with a new Jordan block $J_{r_j-1}(\lambda_j)$. (If $r_j = 1$, we remove the block $J_{r_j}(\lambda_j)$). Note an element x' with this new Jordan form (say viewed in $GL_{n-1}(k)$) has determinant λ^{-1}. However, by scaling and restricting conjugation, we get a new conjugacy class C' of $SL_{n-1}(k)$. Every class C'_i can be extended to the class C_i of $SL_n(k)$ in the obvious way, and rescaling does not effect generation properties.

Lemma 3.6. Let $C_1, \ldots, C_e$ be noncentral conjugacy classes of $SL_n(k)$, where $n \geqslant 3$ and k is an uncountable algebraically closed field. Assume

(i) $\sum_{i=1}^{e} \gamma_i \leqslant n(e-1)$, and
(ii)   it is not the case that $e = 2$, and C_1, C_2 are quadratic.

Finally, assume $C_1', \ldots, C_e'$ are the classes of $SL_{n-1}(k)$ described above. Then $\sum_{i=1}^{e} \gamma_i' \leqslant (n-1)(e-1)$.

> **Proof.**   Pick a class $C = C_i$, and $x \in C$. Assume x has Jordan form $J_{r_1}(\lambda_1) \oplus \ldots \oplus J_{r_t}(\lambda_t)$ on the natural module, and $J_{r_j}(\lambda_j)$ is the Jordan block replaced in x to obtain the Jordan form of elements in $C' = C_i'$. If $r_j = 1$, then $\gamma_i' = \gamma_i - 1$. If $r_j > 1$, then $\gamma_i' = \gamma_i$. In the latter case, all Jordan blocks of x have size greater than one. In particular if $\gamma_i' = \gamma_i$ we have $\gamma_i' \leqslant \frac{n}{2}$ with equality if and only if C is quadratic.
>
> We now argue by induction on the number of classes e. For the base case, assume $e = 2$. Note by condition (i), $\gamma_1 + \gamma_2 \leqslant n$. If either C_1 or C_2 contain a Jordan block of size one, then $\gamma_1' + \gamma_2' \leqslant \gamma_1 + \gamma_2 - 1 \leqslant n - 1$. So assume C_1 and C_2 contain no Jordan blocks of size one. As it is not the case that both C_1 and C_2 are quadratic, we have $\gamma_1' + \gamma_2' \leqslant n - 1$. Now using the inductive hypothesis assume $\sum_{i=1}^{e-1} \gamma_i' \leqslant (e-2)(n-1)$. Then $\sum_{i=1}^{e} \gamma_i' \leqslant (e-1)(n-1)$.   □

Let $\Omega' = C_1' \times \ldots \times C_e'$ be the product of the classes of $SL_{n-1}(k)$ described in the above lemma. We may now complete the proof of the main theorem.

Proof of Theorem 1.1

Assume $n \geqslant 4$. Using the inductive hypothesis, there exists a tuple $\omega' \in \Omega'$ topologically generating $SL_{n-1}(k)$. Extending ω' to a tuple $\omega \in \Omega$, there is an $\omega \in \Omega$ topologically generating a group which contains an isomorphic copy of $SL_{n-1}(k)$. In particular, this implies there is a tuple $\omega \in \Omega$ topologically generating a group which has a composition factor of dimension at least $n - 1$ on the natural module. As ω generates a group acting irreducibly on space of dimension at least $n - 1$, **Lemma 2.5** implies a generic subset of tuples $\Omega_0 \subset \Omega$ share this property. Furthermore **Lemmas 2.5** and **2.15** imply a generic subset of tuples in Ω fix no 1-space or hyperplane (by considering the dual space). Taken in combination, these facts imply generic tuples in Ω act irreducibly on the natural module.

Second, the existence of a tuple $\omega \in \Omega$ topologically generating a group containing a copy of $SL_{n-1}(k)$ implies there is a tuple $\omega \in \Omega$ topologically generating a group N such that any finite index subgroup of N acts irreducibly on a subspace of dimension greater than $n/2$. Recall the free group F_e on e letters has only finitely many subgroups of index d for each divisor d of n. Label these subgroups $H_1(d), \dots, H_t(d)$, and for each $H_i(d)$ choose a collection of generators $w_{i1}, \dots, w_{is}$ for $H_i(d)$. Repeating the argument given in Lemma 2.12, we find the set of tuples $(x_1, \dots, x_e) \in G^e$ such that $\langle w_{i1}(x_1, \dots, x_e), \dots, w_{is}(x_1, \dots, x_e)\rangle$ fixes a common d-dimensional subspace of V is closed in G^e. Then considering each subgroup $H_1(d), \dots, H_t(d)$ for each divisor d of n, we find the set of tuples $(x_1, \dots, x_e) \in \Omega$ having a finite index subgroup acting irreducibly on a subspace of dimension greater than $n/2$ is non-empty and open. It follows there is a tuple $\omega \in \Omega$ generating a group having properties (*i*) and (*ii*) in Lemma 2.11. Applying Lemma 2.12, generic tuples in Ω act primitively on the natural module.

Similarly, using the inductive hypothesis there is a tuple $(x_1, \dots, x_e) \in \Omega$ and word $w \in F_e$ such that $w(x_1, \dots, x_e)$ has infinite order, and a tuple $(x_1, \dots, x_e) \in \Omega$ and word $w \in F_e$ such that $w(x_1, \dots, x_e)$ is strongly regular. Applying Lemmas 2.8 and 2.9, generic tuples in Ω must share both these properties. The finite intersection of generic subsets is a generic set, and over an uncountable field generic sets are non-empty. Hence applying Proposition 2.19, there is a tuple $\omega \in \Omega$ topologically generating G. Finally, applying Lemmas 2.14 and 3.1, we see that the conditions given in the statement of Theorem 1.1 are in fact necessary. $\qquad\square$

4   Variations on the main theorem

In the statement of Theorem 1.1 it is assumed that k is defined over an uncountable algebraically closed field. However once topological generation has been established over uncountable fields, more general results can be recovered under weaker hypotheses on k.

Lemma 4.1.   Let $C_1, \dots, C_e$ be noncentral conjugacy classes of a simply connected simple algebraic group G defined over a field k of characteristic zero. If there is a tuple $\omega \in \Omega$ topologically generating G, then an open dense subset of tuples in Ω topologically generates G.

      Proof.   Let Γ be the set of tuples in Ω that generate a Zariski dense subgroup of G. Note that Γ is defined over k. In characteristic zero, the assumption that k is an algebraically closed field is not required in **Lemma 2.6** and **Remark 2.7** (see **Theorem 4.1** of [2]). So Γ is a nonempty open subset of Ω. □

Lemma 4.2. Let $C_1, \ldots, C_e$ be noncentral conjugacy classes of a simply connected simple algebraic group G, where k is an algebraically closed field of characteristic $p > 0$ that is not algebraic over a finite field. If there is a tuple $\omega \in \Omega$ topologically generating G, then a dense subset of tuples in Ω topologically generate G.

> **Proof.** This follows from **Theorem 2** of [5]. □

Corollary 4.3. Let $C_1, \ldots, C_e$ be a collection of noncentral conjugacy classes of $SL_n(k)$, $n \geq 3$, where k is a field. Assume $\sum_{i=1}^{e} \gamma_i \leq n(e-1)$, and it is not the case that $e = 2$ and C_1, C_2 are quadratic.

(i) If the characteristic of k is zero, then an open dense subset of tuples in Ω topologically generate G.

(ii) If k is an algebraically closed field of positive characteristic that is not algebraic over a finite field, then a dense subset of tuples in Ω topologically generate G.

> **Proof.** **Theorem 1.1** and **Lemma 4.1** imply *(i)*. Statement *(ii)* follows from **Theorem 1.1** and **Lemma 4.2**. □

We may now establish **Theorem 1.2**.

Proof of Theorem 1.2

Corollary 4.3 establishes parts *(ii)* and *(iii)* of **Theorem 1.2**. Taken together, **Lemma 2.10** and **Theorem 1.1** show that part *(i)* holds. □

Finally, let us turn to the case $G = SL_2(k)$, which was excluded in **Theorem 1.1**. It is straightforward to prove an analogous result, although the statement differs slightly. Note if C is a noncentral conjugacy class in $SL_2(k)$, then $\gamma = 1$ and C is quadratic. In particular it is always true that $\sum_{i=1}^{e} \gamma_i \leq n(e-1)$. Hence condition *(i)* of **Theorem 1.1** may be omitted, and condition *(ii)* becomes a statement about classes of involutions modulo the center.

Theorem 4.4. Let C_1, C_2 be noncentral conjugacy classes of $G = SL_2(k)$, where k is an uncountable algebraically closed field. There exists a tuple $\omega \in \Omega$ topologically generating G if and only if it is not the case that C_1 and C_2 are classes of involutions modulo the center.

> **Proof.** Again it suffices to show $\dim \Omega > \dim \Delta + \dim G/M$, where M is any subfield subgroup, or closed maximal positive-

dimensional subgroup of G. First assume M is a subfield subgroup $G(q)$ of G. Any noncentral conjugacy class C of $SL_2(k)$ is two-dimensional. As $\dim(C \cap M) = 0$ and $\dim G/M = 3$, it follows that $\dim \Omega = 4 > \dim \Delta + \dim G/M$.

According to **Theorem 3.2**, each maximal closed positive dimensional subgroup of G is either a Borel subgroup or the normalizer of a maximal torus. To start, assume M is a Borel subgroup. If C_1, C_2 are semisimple, then elements in $C_j \cap M$ have a one dimensional centralizer, and $\dim \Omega > \dim \Delta + \dim G/M$.

A conjugacy class in $SL_2(k)$ is either semisimple or a scalar multiple of a unipotent class. By the above, if C_1 and C_2 are semisimple, then generic tuples in Ω are not contained in a Borel subgroup. So without loss of generality we may assume C_1 is unipotent (up to a scalar). Pick some $x_1 \in C_1$. Since C_1 is a regular unipotent class, x_1 is contained in a unique Borel subgroup. Recall generic tuples in Ω have infinite order. Hence picking an element $x_2 \in C_2$ of infinite order which does not live in this Borel subgroup, we have that $\langle x_1, x_2 \rangle$ topologically generates G. Furthermore since M is parabolic,

$$X = \bigcup_{g \in G} M^g \times M^g$$

is a closed subgroup of $G \times G$, so generic tuples in Ω topologically generate a group living in no conjugate of M.

Finally assume M is the normalizer of a maximal torus. If C is not a class of involutions modulo the center in M, then $\dim (C \cap M) = 0$. Hence $\dim \Omega > \dim \Delta + \dim G/M$, unless C_1, C_2 are both classes of involutions modulo the center. In this latter case, $C_i \subset M^g$ for some $g \in G$. So $\Omega \subset X$, and no tuple in Ω will generate G topologically. $\qquad\square$

The following generalization of **Theorem 4.4** is the analogue of **Theorem 1.1** for $SL_2(k)$.

Theorem 4.5. Let $C_1, \ldots, C_e$ be noncentral conjugacy classes of $G = SL_2(k)$, where k is an uncountable algebraically closed field. There exists a tuple $\omega \in \Omega$ topologically generating G if and only if it is not the case that $e = 2$ and C_1 and C_2 are classes of involutions modulo the center.

Proof.   Assume there is a conjugacy class C in $C_1, \ldots, C_e$ that is not a class of involutions modulo the center. Without loss of generality assume $C \neq C_1$. By **Theorem 4.4**, there exists $x \in C$ and $x_1 \in C_1$ such that $\langle x, x_1 \rangle$ generates G topologically. Furthermore, if

$e = 2$ and C_1 and C_2 are both classes of involutions modulo the center, then no tuple in $C_1 \times C_2$ will topologically generate G. □

5   Applications to generic stabilizers

We now turn our attention to applications of **Theorem 1.1**, the first of which concerns linear representations of algebraic groups. In this section, let G be a simple algebraic group defined over an algebraically closed field k, and let V be a finite dimensional rational kG-module. For any $x \in G$, let

$$V^x = \{v \in V \mid xv = v\}$$

be the vectors in V fixed by x, and

$$V(C) = \{v \in V \mid xv = v \text{ for some } x \in C\}$$

be the vectors fixed by some element in the conjugacy class C. Finally let

$$V^G = \{v \in V \mid xv = v \text{ for all } x \in G\}$$

be the vectors in V fixed by every element of G.

Recall a generic stabilizer is trivial if there exists a non-empty open subvariety $V_0 \subset V$ such that $G_v := \{g \in G \mid gv = v\}$ is trivial for each $v \in V_0$. As discussed in **Section 1**, the cardinality of a generic stabilizer and in particular the question of when a generic stabilizer is trivial arises frequently in the context of invariant theory.

Let $G = SL_n(k)$. We will use **Theorem 1.1** to prove that a generic stabilizer is trivial when $\dim V$ is large enough. The relationship of this result to recent work of Garibaldi-Guralnick [16] [17] [18] and Guralnick-Lawther [25] [26] is discussed in the first section.

To begin, pick some conjugacy class C of G, and assume d elements of C generate G topologically. The first lemma yields a bound on the dimension of $V(C)$ in terms of d and $\dim C$.

Lemma 5.1.   Let $G = SL_n(k)$, where $n \geqslant 3$ and k is an uncountable algebraically closed field. Let V be a finite dimensional rational kG-module. Finally, let C be a conjugacy class of G, and assume d elements of C generate G topologically. Then $\dim V(C) \leqslant \frac{d-1}{d} \cdot \dim V + \dim C$. In particular, $\dim V(C) < \dim V$, when $\dim V > d \cdot \dim C$.

Proof.   Pick $x \in C$ and let $\varphi : G \times V^x \to V$ be the map $(g, v) \mapsto gv$. We claim $im(\varphi) = V(C)$. If $gv \in im(\varphi)$ then $xv = v$

and $gxv = gv$. In particular, $(gxg^{-1})gv = gxv = gv$. So $gv \in V(C)$. Conversely if $v \in V(C)$ there exists a $y \in C$ such that $y = gxg^{-1}$ and $yv = v$. Hence $g^{-1}v \in V^x$ and $\varphi(g, g^{-1}v) = v$, so $v \in im(\varphi)$.

To achieve the desired inequality, we compute a bound on the dimension of a fiber. Pick $v \in V(C) = im(\varphi)$. Then $xv = v$ for some $x \in C$. For any $h \in C_G(x)$, we have $h^{-1}v = h^{-1}xv = xh^{-1}v$. So $h^{-1}v \in V^x$. Clearly $\varphi(h, h^{-1}v) = v$ for all $h \in C_G(x)$. Since $v \in im(\varphi)$ was arbitrary, every fiber has dimension at least $\dim C_G(x)$. In particular,

$$\dim V(C) \leqslant \dim G + \dim V^x - \dim C_G(x).$$

Now let γ be the dimension of the largest eigenspace of a representative of C acting on the natural module. Since d elements of C generate G topologically, it follows from **Theorem 1.1** that $d\gamma \leqslant (d-1)n$. This implies $\dim V^x \leqslant \gamma \leqslant \frac{d-1}{d} \cdot \dim V$. Since $\dim C = \dim G - \dim C_G(x)$, we have

$$\dim V(C) \leqslant \frac{d-1}{d} \cdot \dim V + \dim C.$$

Finally, $\dim V > d \cdot \dim C$ if and only if $\dim V > \frac{d-1}{d} \cdot \dim V + \dim C$, so $\dim V(C) < \dim V$ whenever $\dim V > d \cdot \dim C$. $\quad\square$

Now let $\mathcal{P}$ be the set of noncentral conjugacy classes of G containing elements of prime orders (and containing classes of arbitrary nontrivial unipotent elements in characteristic $p = 0$). The following lemma from the literature allows us to conclude a generic stabilizer is trivial when $\dim V(C) < \dim V$ for all $C \in \mathcal{P}$.

Lemma 5.2.   Let G be a simple algebraic group and V be a finite dimensional rational kG-module such that $V^G = 0$. Let $V(C)$ and $\mathcal{P}$ be as above. If $\dim V(C) < \dim V$ for all $C \in \mathcal{P}$, then a generic stabilizer is trivial.

Proof.   This follows from **Proposition 2.10** in [7] and **Lemma 10.2** in [19]. $\square$

Finally, let $G = SL_n(k)$ and define

$$\mathcal{P}^d := \{C \in \mathcal{P}\,|\, d \text{ elements of } C \text{ are required to generate } G \text{ topologically}\}$$

$$\alpha_d := \max \{\dim C \mid C \in \mathcal{P}^d\}$$

$$\alpha := \max \{\alpha_d \cdot d \mid 2 \leqslant d \leqslant n\}$$

As noted in **Section 1**, it follows from **Theorem 1.1** that if $n \geqslant 3$ and C is a noncentral conjugacy class, then G is topologically generated by n elements in C.

Spencer Gerhardt   70

Now assume $\dim V > \alpha$. Then $\dim V > d \cdot \dim C$, for all $C \in \mathcal{P}$. Applying **Lemma 5.1**, $\dim V(C) < \dim V$ for every $C \in \mathcal{P}$, and hence by **Lemma 5.2** a generic stabilizer is trivial. So to find an upper bound on the dimension of V with a nontrivial generic stabilizer, it suffices to compute an upper bound on α.

Theorem 5.3. Let $G = SL_n(k)$, where $n \geqslant 3$ and k is an algebraically closed field. Let V be a finite dimensional rational kG-module such that $V^G = 0$. If $\dim V > \frac{9}{4}n^2$, then a generic stabilizer is trivial.

Proof. To begin, let k' be an uncountable algebraically closed field extension of k. Note that if a generic stabilizer of $V(k)$ is nontrivial, then the set of points in $V(k)$ with a nontrivial stabilizer is dense. However $V(k)$ is dense in $V(k')$, so a generic stabilizer of $V(k')$ would then also be nontrivial. So without loss of generality we may assume k is an uncountable algebraically closed field. Furthermore, by the reasoning given prior to the statement of the theorem, it is sufficient to show $\alpha \leqslant \frac{9}{4}n^2$.

First we consider $\mathcal{P}^2$. Let C be a regular semisimple conjugacy class in G. Then C is not quadratic and by **Theorem 1.1**, $C \in \mathcal{P}^2$. Regular semisimple classes are conjugacy classes of maximal dimension in G, so $\dim C = \alpha_2 = n^2 - n$.

For $d > 2$, pick $C \in \mathcal{P}^d$ and $x \in C$. By **Theorem 1.1**, $\gamma \geqslant \frac{d-2}{d-1} \cdot n$. It follows that $C_G(x)$ must contain a copy of $GL_\beta(k)$ where $\beta = \lceil \frac{n(d-2)}{d-1} \rceil$. It then follows from **Proposition 2.9** in [4] that $\alpha_d \leqslant n^2 - \beta^2 - (n - \beta)$, noting that $s = n - \beta$ in the notation of the reference.

Finally, we compute an upper bound for $\alpha = \max \{\alpha_d \cdot d \mid 2 \leqslant d \leqslant n\}$. Note $\alpha_2 = n^2 - n$, and for $3 \leqslant d \leqslant n$:

$$\alpha_d \leqslant n^2 - \beta^2 - (n - \beta) < n^2 - \beta^2 \leqslant n^2 - \frac{n^2(d-2)^2}{(d-1)^2}.$$

Furthermore,

$$d\left(n^2 - \frac{n^2(d-2)^2}{(d-1)^2}\right) = n^2\frac{d(2d-3)}{(d-1)^2}$$

and $\frac{d(2d-3)}{(d-1)^2} \leqslant \frac{9}{4}$ for $3 \leqslant d \leqslant n$. So

$$\alpha \leqslant \max \left(2(n^2 - n), n^2\frac{d(2d-3)}{(d-1)^2}\right) \leqslant \frac{9}{4}n^2.$$

$\square$

Note that it is not claimed the above bound is sharp. However it is on the order of the best possible bound, as discussed in **Section 1**. For instance, there are examples where $G = SL_n(k)$, $V^G = 0$, $\dim V > n^2$ and a generic stabilizer is nontrivial.

6 Applications to random generation

We now turn to a final application of topological generation, concerning the random generation of finite simple groups of Lie type. Let H be a finite group and $I_t(H)$ be the elements of order t in H. Assume H contains elements of prime orders r and s, and let

$$P_{r,s}(H) := \frac{|\{(x, y) \in I_r(H) \times I_s(H) : \langle x, y \rangle = H\}|}{|I_r(H) \times I_s(H)|}$$

be the probability that two random elements of orders r and s generate H. Now let H_i be a sequence of finite simple groups of Lie type (of some fixed type) and assume $|H_i| \to \infty$ as $i \to \infty$. (Here the rank of the groups H_i may vary, or the field they are defined over, or both). If $P_{r,s}(H_i) \to 1$ as $i \to \infty$, then groups of the relevant type are said to have *random (r, s)-generation*.

The following result of Liebeck-Shalev [48] shows that finite simple classical groups have random (r, s)-generation when the rank of the group is large enough (depending on the primes r and s). In fact, explicit bounds on $f(r, s)$ are determined in [59].

Theorem 6.1. Let (r, s) be primes with $(r, s) \neq (2, 2)$. There exists a positive integer $f(r, s)$ such that if H is a finite simple classical group of rank at least $f(r, s)$, then

$$P_{r,s}(H) \to 1 \text{ as } |H| \to \infty.$$

The methods used to prove **Theorem 6.1** in [48] are probabilistic in nature. We will use different tools, namely **Theorem 1.1** and some algebraic geometry to establish the random (r, s)-generation of linear and unitary groups, where the rank is fixed and r, s are independent of the rank. This gives a strengthening of **Theorem 6.1** in these cases.

Let G be a simply connected simple algebraic group defined over the algebraic closure k of a prime field. Let F or $F_q : G \to G$ be a Steinberg endomorphism, G^F or $G(q)$ be the fixed points of F on G, and $E(q) := E \cap G(q)$ for any subset E of G. In particular, if $G = SL_n(k)$ we have $G(q) = SL_n(q)$ or $SU_n(q)$. Finally, if C_1, C_2 are conjugacy classes of G such that $\Omega(q) := C_1(q) \times C_2(q) \neq \emptyset$, let $P_{\Omega(q)}(G(q))$ be the probability that a random pair in $\Omega(q)$ generates $G(q)$.

By **Lemma 2.2** of [27], there exists a finite collection $\mathcal{U} = \{V_1, \ldots, V_d\}$ of finite-dimensional irreducible rational kG-modules such that every proper closed subgroup of G that acts irreducibly on each $V_i \in \mathcal{U}$ is conjugate to some subfield subgroup $G(q)$ of G.

In particular, for some fixed subfield subgroup $G(q_0)$ of G, there is a set W_{q_0} such that both of the following hold:

(i) W_{q_0} consists of the tuples $(x_1, \ldots, x_e) \in G^e$ that act irreducibly on each module $V_i \in \mathcal{U}$.

(ii)   Each tuple in W_{q_0} generates either a Zariski dense subgroup of G, or a conjugate of some subfield subgroup $G(q)$ of G, with $G(q) \geqslant G(q_0)$.

By **Lemma 2.4** of [27], W_{q_0} is defined over $\mathbb{F}_p$, and forms an open dense subset of G^e. Counting $\mathbb{F}_q$-points on $\Omega \cap W_{q_0}$, the following theorem relates the existence of a tuple $\omega \in \Omega$ topologically generating G to random tuples in $\Omega(q)$ generating the corresponding finite group of Lie type $G(q)$.

Theorem 6.2.   Let G be a simply connected simple algebraic group defined over an algebraically closed field k of characteristic $p > 0$. Let C_1 and C_2 be conjugacy classes of G, $\Omega = C_1 \times C_2$, and Q be the set of powers $q = p^a$ such that $\Omega(q) =: C_1(q) \times C_2(q) \neq \emptyset$. The following are equivalent:

(a)   If k is uncountable, there exists a tuple $\omega \in \Omega$ topologically generating G.

(b)   If k is uncountable, a generic subset of tuples in Ω topologically generate G.

(c)   If k is not algebraic over a finite field, there exists a tuple $\omega \in \Omega$ topologically generating G.

(d)   If k is not algebraic over a finite field, a dense subset of tuples in Ω topologically generate G.

(e)   If $q \in Q$ is sufficiently large, then there exists a tuple $(x_1, x_2) \in \Omega(q)$ such that $\langle x_1, x_2 \rangle = G(q)$.

(f) $\lim_{q \in Q, q \to \infty} P_{\Omega(q)}(G(q)) = 1$.

> **Proof.**   The equivalence of $(a) - (d)$ follows from **Lemmas 2.10, 4.1** and **4.2** above. The equivalence of (e) and (f) is **Corollary 5** in [27]. Hence it suffices to show (a) is equivalent to (e).
>
> To see that (e) implies (a), assume k is an uncountable algebraically closed field. For any word $w(x, y)$, define a trace map
>
> $$tr_w : C_1 \times C_2 \to k, \quad (x_1, x_2) \mapsto tr(w(x_1, x_2))$$
>
> where the natural representation is taken for classical groups and the adjoint representation is considered for exceptional groups.

The proof of **Theorem 3.1** in [27] shows if there is a tuple $(x_1, x_2) \in \Omega(q)$ such that $\langle x_1, x_2 \rangle = G(q)$ for sufficiently large q, there is a fixed word $w(x, y)$ such that $tr_w(x, y)$ is non-constant on $C_1 \times C_2$. Furthermore by the proof of **Corollary 2.8** *(ii)* in [27], the trace map is a morphism mapping into a 1-dimensional variety k, so its image contains an open subset of its closure. Hence over any algebraically closed field k, its image attains all but finitely many values of k. In particular, there is a tuple $(x_1, x_2) \in \Omega$ such that $tr_w(x_1, x_2)$ is not in the algebraic closure of $\mathbb{F}_p$, and hence $w(x_1, x_2)$ has infinite order. It then follows from **Lemma 2.9** that generic tuples in Ω generate a group of infinite order. Since for some sufficiently large q there is a tuple $(x_1, x_2) \in \Omega(q)$ such that $\langle x_1, x_2 \rangle = G(q)$, $W_{q_0} \cap \Omega$ is a nonempty open subset of Ω. The intersection of generic and open subsets is nonempty, so some tuple $\omega \in \Omega$ topologically generates G.

Finally for (a) implies (e), assume there is a tuple $\omega \in \Omega$ topologically generating G, and $\dim \Omega = s$. Then $W_{q_0} \cap \Omega \neq \emptyset$, and hence forms an open dense subset of Ω. It follows $\Omega \backslash W_{q_0}$ is a proper closed subvariety of Ω. Applying the Lang-Weil theorem [43] there is a positive constant c such that for all sufficiently large $q \in Q$, $|\Omega(q)| > cq^s$. Hence there is a positive constant c_1 such that for sufficiently large q, $|\Omega(q) \backslash W_{q_0}| \leqslant c_1 q^{s-1}$. In particular for sufficiently large q, almost all tuples of $\Omega(q)$ are contained in $\Omega(q) \cap W_{q_0}$.

Next we count the proportion of pairs in $C_1(q) \times C_2(q)$ which are contained in some conjugate of $G(q_1) \times G(q_1)$, where $G(q_1)$ is a proper subfield subgroup of $G(q)$. By **Lemma 2.5** of [27], $s = \dim \Omega = \dim G + e$ for some $e \geqslant 1$. Now assume $q = p^a$, $q_1 = q^{\frac{1}{b}}$, and $d = \dim G$. There exist positive constants c_2 and c_3 such that for large enough q, $|\Omega(q_1)| \leqslant c_2 q^{\frac{s}{b}}$, and $|G(q) : G(q_1)| \leqslant c_3 q^{d(1-\frac{1}{b})}$. Hence there is a positive constant c_4 such that for sufficiently large q, the proportion of pairs in $C_1(q) \times C_2(q)$ which are contained in some conjugate of $G(q_1) \times G(q_1)$ is bounded by

$$\frac{|\Omega(q_1)||G(q) : G(q_1)|}{|\Omega(q)|} \leqslant c_4 \left(\frac{q^{\frac{s}{b}} \cdot q^{d(1-\frac{1}{b})}}{q^s} \right) = c_4 q^{e(\frac{1}{b}-1)}$$

Summing over all positive divisors b of a, we find the probability that a tuple in $\Omega(q)$ is contained in a proper subfield subgroup of $G(q)$ is at most $c_4 \sum_{b|a} q^{e(\frac{1}{b}-1)}$ and hence is bounded above by $O(q^{-\frac{e}{2}})$. So for any sufficiently large q, there is a tuple $(x_1, x_2) \in \Omega(q) \cap W_{q_0}$ not contained in any proper subfield subgroup $G(q_1)$ of $G(q)$. As $\langle x_1, x_2 \rangle \subset G(q)$ and tuples in W_{q_0} generate a group containing a subfield subgroup of G, it follows that $\langle x_1, x_2 \rangle = G(q)$. □

Remark 6.3. Note while the results of Liebeck-Shalev [48] are stated for finite simple classical groups, G^F itself is not typically simple. This small discrepancy is easily resolved. If G is a simply connected simple algebraic group, then $S = G^F/Z(G^F)$ is (almost always) a finite simple group. More generally, $[G^F, G^F]/Z([G^F, G^F])$ is simple except for a very small number of cases. Letting $\varphi : G^F \to S$ be the natural projection map, the center of G^F is contained in the Frattini subgroup of G^F, and hence tuples $(x_1, x_2) \in \Omega(q)$ generate G^F if and only if the corresponding pair $(\overline{x_1}, \overline{x_2})$ generates S. (Of course, the orders of the elements may be different in S).

An element $\overline{x} \in S$ has prime order r if and only if the corresponding element $x \in G^F$ has prime power order $q = r^a$ for some $a \geqslant 1$. In particular, elements $\overline{x_1}, \overline{x_2} \in S$ with prime orders t and u generate S if and only if the corresponding elements x_1 and x_2 with prime power orders $r = t^a$ and $s = u^b$ generate G. Hence to prove random (t, u)-generation for primes t and u in the simple group S, it suffices to prove random (r, s)-generation in G^F for prime powers r and s.

We now give the condition that will be used to establish the desired random (r, s)-generation result. To begin, fix prime powers r and s, and let G be a simply connected simple algebraic group. We say the pair of conjugacy classes C' and D' in G are *bad* if the following hold:

(i) C' and D' are classes of elements of orders r and s in G and,
(ii) $\Omega'(q) = C'(q) \times D'(q) \neq \emptyset$ and,
(iii) there is no tuple $\omega' \in \Omega'$ topologically generating G.

Similarly we say the pair of conjugacy classes C and D in G are *good* if the following hold:

(i) C and D are classes of elements of orders r and s in G and,
(ii) $\Omega(q) = C(q) \times D(q) \neq \emptyset$ and,
(iii) there is a tuple $\omega \in \Omega$ topologically generating G and,
(iv)   for any bad classes C' and D' of G, $\dim \Omega' < \dim \Omega$.

Lemma 6.4. Let G be a simply connected simple algebraic group defined over an uncountable algebraically closed field of characteristic $p > 0$. Let Q be the set of powers $q = p^a$ such that $G(q)$ contains elements of prime power orders r and s. If good classes exist for each $q \in Q$, then

$$\lim_{q \in Q, q \to \infty} P_{r,s}(G(q)) = 1.$$

Proof.   Fix $q \in Q$, and let C_1, C_2 be good classes corresponding to q. Next let $A_1, \ldots, A_l$ be all conjugacy classes of order r elements in G such that $A_i(q) \neq \emptyset$, and $A = \bigcup_{i=1}^{l} A_i$. Similarly, let $B_1, \ldots, B_m$ be all conjugacy classes of order s elements in G such that $B_i(q) \neq \emptyset$, and $B = \bigcup_{i=1}^{m} B_i$. Finally let $X = A \times B$, and $\Omega = C_1 \times C_2$.

Since C_1 and C_2 are good classes, $\dim \Omega = \dim X$. Hence there is a positive constant c such that for sufficiently large q, $|X(q)| < c|\Omega(q)|$. It follows that

$$\lim_{q \in Q, q \to \infty} P_{\Omega(q)}(G(q)) = \lim_{q \in Q, q \to \infty} P_{X(q)}(G(q))$$

where $P_{X(q)}(G(q))$ and $P_{\Omega(q)}(G(q))$ are the probabilities that random tuples in $X(q)$ and $\Omega(q)$ generate $G(q)$.

Now for each $q_i \in Q$, define Ω_i and X_i as above by choosing appropriate good classes. Let Q_i be the set of powers $q = p^a$ such that $\Omega_i(q) \neq \emptyset$. **Theorem 6.2** implies

$$\lim_{q \in Q_i, q \to \infty} P_{\Omega_i(q)}(G(q)) = \lim_{q \in Q_i, q \to \infty} P_{X_i(q)}(G(q)) = 1.$$

Since this holds for each i, it follows that

$$\lim_{q \in Q, i, q \to \infty} P_{X_i(q)}(G(q)) = \lim_{q \in Q, q \to \infty} P_{r,s}(G(q)) = 1.$$

$\square$

Combined with **Lemma 6.4**, the following result completes the proof of **Theorem 1.4**.

Theorem 6.5. Let $G = SL_n(k)$, $n \geq 3$, where k is an uncountable algebraically closed field of positive characteristic. Assume $(r, s) \neq (2, 2)$ are prime powers, and $F_q : G \to G$ is a Steinberg endomorphism such that $G(q)$ contains elements of orders r and s. Then G contains a pair of good classes of elements of orders r and s.

Proof. First assume $G(q) = SL_n(q)$ and recall that

$$|G(q)| = q^{\frac{n(n-1)}{2}}(q^2 - 1)(q^3 - 1) \cdots (q^n - 1).$$

Without loss of generality, assume $r > 2$. Pick any maximal dimensional conjugacy class C of order r elements in G, with $C(q) \neq \emptyset$. Note that C is not quadratic. Let γ_1 be the dimension of the largest eigenspace of C. We claim $\gamma_1 < \frac{n}{2}$.

If $r \mid q$ (so that the classes under consideration are unipotent) or $r \mid q - 1$, then all classes of order r elements are defined in $G(q)$, and $\gamma_1 \leq \lceil \frac{n}{3} \rceil$. So assume $r \nmid q, q - 1$, and pick $x \in C(q)$. Consider the action of x on the natural module V and define

$$l = \min \{j : r \mid q^j - 1 \text{ and } 1 \leqslant j \leqslant n\},$$

$$m = \max \{j : r \mid q^j - 1 \text{ and } 1 \leqslant j \leqslant n\}.$$

Note x is semisimple, acts reducibly on a space of dimension at least $n - m$ with $n - m < \frac{n}{2}$ (since r is a prime power, any element of order r in $G(q)$ has a fixed space of dimension at least $n - m$), and has at most m/l irreducible composition factors of dimension l, with $1 < l \leqslant n$. A maximal dimensional conjugacy class of order r elements in G such that $C(q) \neq \emptyset$ does not have $m/2$ irreducible composition factors of dimension two. It follows that $\gamma_1 \leqslant \max(n - m, \frac{n}{2} - 1) < \frac{n}{2}$, and C is not quadratic.

Next let D be any maximal dimensional class of order s elements in G such that $D(q) \neq \emptyset$. Let γ_2 be the dimension of the largest eigenspace of a representative in D. If $s > 2$, then by the above we immediately have $\gamma_1 + \gamma_2 \leqslant n$. If $s = 2$, and D is a maximal dimensional conjugacy class of involutions, then $\gamma_2 = \lceil \frac{n}{2} \rceil$. Again we have $\gamma_1 + \gamma_2 \leqslant n$. In both cases either C or D is not quadratic, and hence it follows from **Theorem 1.1** that there is a tuple $\omega \in \Omega$ topologically generating G.

So for any prime powers $(r, s) \neq (2, 2)$ and any maximal dimensional conjugacy classes C and D of G containing elements of orders r and s, respectively, with the property $\Omega(q) \neq \emptyset$ for $\Omega = C \times D$, there is a tuple $\omega \in \Omega$ topologically generating G. Hence for each Steinberg endomorphism $F_q : G \to G$ defining $G(q) = SL_n(q)$, good classes of prime powers $(r, s) \neq (2, 2)$ exist.

Finally, assume $F_q : G \to G$ is a Steinberg automorphism such that $G(q) = SU_n(q)$. Then

$$|G(q)| = q^{\frac{n(n-1)}{2}}(q^2 - 1)(q^3 + 1) \cdots (q^n - (-1)^n).$$

However, if $r \mid q^l \pm 1$ and $r \nmid q^j \pm 1$ for $l < j \leqslant n$, then $l \leqslant \frac{n}{2} - 1$. We may repeat the reasoning above to show good classes C, D exist for prime powers $(r, s) \neq (2, 2)$. $\qquad \square$

Remark 6.6. Note if $G = SL_2(k)$, then by **Theorem 4.5** bad classes do not exist unless C_1 and C_2 are both classes of involutions modulo the center. Hence it follows from **Theorem 6.5** and **Remark 6.3** that $PSL_2(q)$ has random (r, s)-generation for primes $(r, s) \neq (2, 2)$.

Acknowledgements

This research was partially supported by NSF grant DMS-1302886. The author would like to thank Robert Guralnick for many helpful comments during the development of this paper. In addition, thanks are due to Gunter Malle for numerous valuable remarks on an earlier draft of this article, and to the anonymous referee whose careful reading led to several organizational changes and improvements in the text.

References

[1]   M. Aschbacher, *On the maximal subgroups of the finite classical groups*, Invent Math (1984), 76: 469–514.

[2]   E. Breuillard, B. Green, R. Guralnick, and T. Tao, *Strongly dense free subgroups of semisimple algebraic groups*, Isr. J. Math 192 (2012), 347–379.

[3]   P. Brosnan, Z. Reichstein, and A. Vistoli, *Essential dimension, spinor groups, and quadratic forms*, Ann. of Math. 2, 171 (2010), 533–544.

[4]   T. Burness, *Fixed point spaces in actions of classical algebraic groups*, J. Group Theory 7 (2004), no. 3, 311–346.

[5]   T. Burness, S. Gerhardt and R. Guralnick, *Topological generation of exceptional algebraic groups*, Adv. Math. 369 (2020).

[6]   T. Burness, S. Gerhardt and R. Guralnick, *Topological generation of classical algebraic groups*, in preparation to appear in the Journal of the European Mathematical Society.

[7]   T. Burness, R. Guralnick and J. Saxl, *On base sizes for algebraic groups*, Journal of the European Mathematical Society, 19 (2013), 2269–2341.

[8]   M. Chiodo, *Finitely annihilated groups*, Bull. Austral. Math. Soc. 90, (2014), 404–417.

[9]   W. Crawley-Boevey, *Indecomposable parabolic bundles and the existence of matrices in prescribed conjugacy class closures with product equal to the identity*, Publications mathématiques de l'IHES, 100 (2004), 171–207.

[10]   W. Crawley-Boevey, *Quiver algebras, weighted projective lines, and the Deligne-Simpson problem*, International Congress of Mathematicians. Vol. II, Eur. Math. Soc., Zurich, (2006), 117–129.

[11]   W. Crawley-Boevey and P. Shaw, *Multiplicative preprojective algebras, middle convolution and the Deligne-Simpson problem*, Adv. Math. 201 (2006), 180–208.

[12]   J. D. Dixon, *The probability of generating the symmetric group*, Math. Z. 110 (1969), 199–205.

[13]   A. Eisenmann and N. Monod, *Normal generation of locally compact groups*. Bull. Lond. Math. Soc. 45 (2013), no. 4, 734–738.

[14]   A.G. Élašvili, *Canonical form and stationary subalgebras of points in general position for simple linear Lie groups* (Russian), Functional. Anal. i Prilozen. 6 (1972), 51–62.

[15]   A.G. Élašvili, *Stationary subalgebras of points of general position for irreducible linear Lie groups* (Russian), Functional. Anal. i Prilozen. 6 (1972), 65–78.

[16]   R. Guralnick, *Generically free representations I: large representations*, Algebra Number Theory 14, vol. 6 (2020), 1577–1611.

[17]   S. Garibaldi and R. Guralnick, *Generically free representations II: irreducible representations*, Transformation Groups 25 (2020), 793–817.

[18]   S. Garibaldi and R. Guralnick, *Generically free representations III: exceptionally bad characteristic*, Transformation Groups 25 (2020), 819–841.

[19]   S. Garibaldi and R. Guralnick, *Simple groups stabilizing polynomials*, Forum of Mathematics, Pi, 3, E3, 2015.

[20]   T. Gelander, *Convergence groups are not invariably generated*, Int. Math. Res. Not. (2015), 9806–9814.

[21]   T. Gelander, G. Golan and K. Juschenko, *Invariable generation of Thompson groups*, J. Algebra 478, (2017), 261–270.

[22]   T. Gelander and C. Meiri, *The congruence subgroup property does not imply invariable generation*, Int. Math. Res. Not. (2017), 4625–4638.

[23]   A. Grothendieck, *Élements de géométrie algébrique (rédigés avec la collaboration de J. Dieudonné): Étude locale des schémas et des morphismes de schémas*, III. Inst. Hautes Etudes Sci. Publ. Math. No. 28, 1966.

[24]   R. Guralnick, *Some applications of subgroup structure to probabilistic generation and covers of curves*, in Algebraic groups and their representations, Cambridge (1997), 301–320, NATO Adv. Sci. Inst. Ser. C Math. Phys. Sci., 517, Kluwer Acad. Publ., Dordrecht.

[25]   R. Guralnick and R. Lawther, *Generic stabilizers in actions of simple algebraic groups I: modules and first Grassmannian varieties*, preprint, arXiv:1904.13375.

[26]   R. Guralnick and R. Lawther, *Generic stabilizers in actions of simple algebraic groups II: higher Grassmannian varieties*, preprint, arXiv:1904.13382.

[27]   R. Guralnick, M.W. Liebeck, F. Lübeck, and A. Shalev, *Zero-one generation laws for finite simple groups*, Proceedings of the American Mathematical Society, 147.

[28]   R. Guralnick and J. Saxl, *Generation of finite almost simple groups by conjugates*, J. Algebra 268 (2003), 519–571.

[29]   R. Guralnick and P. Tiep, *Decompositions of small tensor powers and Larsen's conjecture*, Represent. Theory 9 (2005), 138–208.

[30]   J. Hall, M. Liebeck, and G. Seitz, *Generators for finite simple groups, with applications to linear groups*, Q. J. Math 43 (1992), 441–458.

[31] D. Johnson. *Homomorphs of knot groups*, Proc. Amer. Math. Soc., vol. 78, no. 1, (1980), 135–138.

[32] W.M. Kantor and A. Lubotzky. *The probability of generating a finite classical group*, Geom. Dedicata 36 (1990), 67–87.

[33] W.M. Kantor, A. Lubotzky and A. Shalev *Invariable generation of infinite groups*, J. Algebra 421 (2015), 296–310.

[34] W. M. Kantor, A. Lubotzky and A. Shalev, *Invariable generation and the Chebotarev invariant of a finite group*, J. Algebra 348 (2011), 302–314.

[35] N. Karpenko, *Canonical dimension*, Proceedings of the International Congress of Mathematicians 2010, World Scientific, 2010.

[36] N. M. Katz, *Rigid local systems*, Princeton University Press, Princeton, NJ, 1996.

[37] S. Kim, *Normal generation of line bundles on multiple coverings*. J. Algebra 323 (2010), no. 9, 2337–2352.

[38] V. P. Kostov, *On the existence of monodromy groups of Fuchsian systems on Riemann's sphere with unipotent generators*, J. Dynam. Control Systems 2 (1996), 125–155.

[39] V. P. Kostov, *On the Deligne-Simpson problem*, C. R. Acad. Sci. Paris Sér. I Math. 329 (1999), 657–662.

[40] V. P. Kostov, *On some aspects of the Deligne-Simpson problem*, J. Dynam. Control Systems 9 (2003), 393–436.

[41] V. P. Kostov, *The Deligne-Simpson problem - a survey*, J. Algebra 281 (2004), 83–108.

[42] V.P. Kostov, *Some examples related to the Deligne-Simpson problem*, Second International Conference on Geometry, Integrability and Quantization, June 7–15, 2000, Varna, Bulgaria Ivaïlo M. Mladenov and Gregory L. Naber, Editors Coral Press, Sofia 2001.

[43] S. Lang and A. Weil, *Number of points of varieties in finite fields*, Amer. J. Math. 76 (1954), 819–827.

[44] R. Lawther, M.W. Liebeck and G.M. Seitz, *Fixed point spaces in actions of exceptional algebraic groups*, Pacific J. Math. 205 (2002), 339–391.

[45] M.W. Liebeck and G.M. Seitz, *On the subgroup structure of the classical groups*, Invent. Math. 134 (1998), 427–453.

[46] M.W. Liebeck and G.M. Seitz, *The maximal subgroups of positive dimension in exceptional algebraic groups*, Mem. Amer. Math. Soc. 802, 2004.

[47] M.W. Liebeck and A. Shalev. *The probability of generating a finite simple group*, Geom. Dedicata 56 (1995), 103–113.

[48] M.W. Liebeck and A. Shalev, *Random (r,s)-generation of finite classical groups*, Bull. London Math. Soc. 34 (2002), 185–188.

[49] M.W. Liebeck and A. Shalev, *Classical groups, probabilistic methods, and the 2, 3 -generation problem*, Ann. of Math 144 (1996), 77–125.

[50] R. Lötscher, M. MacDonald, A. Meyer, and Z. Reichstein, *Essential dimension of algebraic tori*, J. Reine Angew. Math. 677 (2013), 1–13.

[51] F. Lübeck, *Small degree representations in defining characteristic*, LMS J. Comput. Math 56 (2001), 135–169.

[52] F. Lübeck and G. Malle, *(2,3)-Generation of Exceptional Groups*, Journal of the London Mathematical Society, 59, (1999), 109–122.

[53] D. Osin and A. Thom, *Normal generation and 2-Betti numbers of groups*. Math. Ann. 355 (2013), 1331–1347.

[54] A.M. Popov, *Finite stationary subgroups in general position of simple linear Lie groups* (Russian), Trudy Moskov. Mat. Obshch. 48 (1985), 7–59.

[55] A.M. Popov, *Finite isotropy subgroups in general position of irreducible semisimple linear Lie groups* (Russian), Trudy Moskov. Mat. Obshch.50 (1987), 209–248.

[56] V.L. Popov and E.B. Vinberg, *Invariant theory*, in Encyclopaedia of Mathematical Sciences series, vol. 55, Algebraic Geometry IV, Springer-Verlag, Berlin, 1994.

[57] R. W. Richardson, *Principal orbit types for algebraic transformation spaces in characteristic zero*, Invent. Math. 16 (1972), 6–14.

[58] C. T. Simpson, *Products of Matrices*, Differential geometry, global analysis, and topology (Halifax, NS, 1990), Canadian Math. Soc. Conf. Proc. 12 (1992), Amer. Math. Soc., Providence, RI, 1991, pp. 157–185.

[59] M. Stavrides, *On the random generation of finite simple classical groups*, Comm. Algebra 32 (2004), 4273–4283.

[60] A. Thom, *A note on normal generation and generation of groups*. Commun. Math. 23 (2015), no. 1, 1–11.

[61] J. Wiegold, *Transitive groups with fixed-point-free permutations*, Arch. Math. (Basel) 27 (1976), 473–475.

[62] J. Wiegold, *Transitive groups with fixed-point-free permutations II*, Arch. Math. (Basel) 29 (1977), 571–573.

Rodarte
Zerlina costume design for Los Angeles Philharmonic
    production of *Don Giovanni*, 2012
Pencil on paper
Courtesy of the artists

Reverse Perspectives

85

I

Grothendieck noted two distinct approaches in mathematics. One is the hammer and chisel, applying direct force in gainful locations. The second is an indirect method, where the object of inquiry is subsumed in a larger theory, so in time the "nourishing flesh" reveals itself effortlessly and with little force.[1]

> I can illustrate that second approach with the same image of a nut to be opened. The first analogy that came to my mind is of immersing the nut in some softening liquid, and why not simply water? From time to time you rub so the liquid penetrates better, and otherwise you let time pass. The shell becomes more flexible through weeks and months—when the time is ripe, hand pressure is enough, the shell opens like a perfectly ripened avocado!

> A different image came to me a few weeks ago. The unknown thing to be known appeared to me as some stretch of earth or hard marl, resisting penetration. . . the sea advances insensibly in silence, nothing seems to happen, nothing moves, the water is so far off you hardly hear it . . . yet finally it surrounds the resistant substance.[2]

Reflecting on how to introduce this varied collection[3] it occurred to me that these pieces, like propositions in Grothendieck's "vast theory," arise naturally and with little force, but only out of a larger context of performance and interpretation. One developed over the course of a life, and which I've had a unique opportunity to observe. (As a fortuitous coincidence, this anthology spans the exact period I've known Charles, and the cover photo was taken the night we met.) Indeed, one of the distinguishing features of Charles's performance style is his insistence on, and almost philosophical concern for, a larger framework of transmission that includes performance, recording, writing, acoustics, architecture, visual presentation, etc. Most surprisingly, this framework is somehow invariably applicable, even when technical issues are in question. It is this ambient theory, which permeates these recordings, that I seek to describe.

II

My most intensive interactions with Charles occurred in 2005–06, when I moved from Amsterdam to San Diego to work with him. This was an incredibly fruitful and productive period for him—putting together "Waking States," a concert series that in some sense served as the precursor to this anthology, working with La Monte Young on the premiere of a new version of *Trio for Strings* (1958), beginning an intensive period of collaboration with Éliane Radigue and Alvin Lucier, in addition to being part of a family with three young children, overseeing a major home renovation, and keeping up a demanding performance and teaching schedule.

Our meetings typically started before and continued after the family dinner at their Ocean Beach home, a compound consisting of a modest and anonymous cinder block structure, a beautifully detailed 1920s bait shop, which had been transferred to the front of the property, and a two-story glass parallelepiped in the process of being slid between the two. Often, time was spent working through ideas and areas of common interest—notions of perspective, theories of performance and interpretation, Young's *Compositions 1960* and the idea of tuning across his music oeuvre, the work of Piero della Francesca and Albrecht Durer—and, perhaps foremost, listening to classical music.

One striking feature of these conversations was an underlying viewpoint that was both unfamiliar and never fully stated, perhaps because even more improbably it was simply a lived attitude. It is best described by example.

Through his engagement with Lucier, Charles became interested in the idea that acoustical beating travels in an enclosed space according to a certain orientation. We agreed to meet at University of California, San Diego, to discuss this idea. Having some scientific background, I tried to find a reference that would either confirm this theory or call it into question. I can't remember what I discovered, but upon arriving in the basement of Mandeville Hall we sat in various locations on opposite ends of a large dungeonlike classroom listening to pairs of sine tones for what seemed to be a very long time, attempting to describe what we were hearing. Whether any scientific conclusion could be drawn from this experiment is debatable, but the principle itself was striking: the confidence to reconstruct the world (or a world?) through lived experience, even when a physical theory was in question. (I later had a similar experience studying with Young, and I suspect it is precisely this disposition that drew them indelibly together.)

This phenomenon repeats itself on several occasions. For thirty years Young had sought a global tuning for *Trio for Strings* that could clarify the notion of harmonic identity lying at the center of the composition. Through intensive discussions with Young, Charles discovered a remarkably simple and elegant solution to this problem, combining the twelve-tone structure of the piece and the fundamental notion of a Dream Chord.

Shortly after arriving in San Diego, Charles described the tuning to me. Its essential correctness was apparent, and I wrote a note formalizing the underlying reasoning and showing it was, in an appropriate sense, the unique and optimal solution to the problem. (However, the existence of such a tuning at all reflects a remarkable feat of compositional intuition on Young's part.) In my mind I was transcribing a process of thought, but after a minutes' discussion it became apparent that my attempted translation in fact bore little resemblance to his conception of the tuning, and the combinatorial possibilities and counterfactual situations it explored were not of particular interest to him. Instead, the idea emerged through his careful listening to the chords appearing in exposition that begins the piece, his long engagement with Young's music, and his deep awareness of the audible structure of the harmonic series. For Charles, the tuning was first and foremost an auditory solution, and one whose beauty arose simply out of experience.

As a final example, I recall several trips to Los Angeles before moving here, in particular one occasion to see a screening of *La Région Centrale* (1971) at the Egyptian Theatre. Often these trips involved detours to various points of interest, and Charles would navigate via a system of personal landmarks; these guideposts were invariably locations of classical music lessons or recitals he had as a child coming up from Laguna Beach.

That one could reconstruct the vast terrain of Los Angeles through childhood memories of being driven around in the family car seemed indicative of both an exceptional level of awareness of and engagement with the surrounding environment. The fact that something as ultimately objective as arriving at a given location could arise out of an intensely subjective awareness of the past provides insight into his style as a performer. I was later reminded of this phenomenon when I was singing raga with Young and Zazeela, the act of tuning to higher partials of the tambura being a similarly nonlinear process of navigation filtered through memory and experience. To this day Charles slowly provides directions to his current location, even when it must be clear that everyone types addresses into their phones. (His now-grown daughters have commented humorously on this habit.)

Of course, at some point experience merges with autobiography, and one cannot help but sense a distinctive air of longing in the tracks on *Performances & Recordings: 1998–2018*. Of introspection and of time passing, tuning itself being an exemplary model of this process. In spirit, it is similar to how Brouwer described mathematics:

> The fullest constructional beauty is the *introspective beauty of mathematics*, where . . . the basic intuition [of time passing] is left to free unfolding. This unfolding is not bound to the exterior world, and thereby to finiteness and responsibility; consequently its introspective harmonies can attain any degree or richness and clearness.[4]

That intonation itself could be explored as a form of lived experience, or as expression of lived experience, is a distinguishing feature of both his style and repertoire.

After one unusually long period of silence I ran into Charles in New York, on the occasion of his performance of Terry Jennings's *Piece for Cello and Saxophone* (1960) at Blank Forms. We met after the concert, and there was much to catch up on. However, he immediately started describing a theory of icon painting he recently discovered: Pavel Florensky's reverse perspective. This theory begins from the simple observation that in many iconographic works objects farther in the distance appear larger than objects nearer by. From this a more speculative proposal was developed, that of a projective space where the point at infinity was taken to be the position of the observer.

Although the meaning of this seeming reversal is still much debated, the notion that the true vanishing point is simply one's own subjective experience seemed to illuminate many years of conversation, and in fact the atmosphere and content of that evening's performance. A romantic space, consisting of a spare collection of chords and evolving melodic patterns, modulating in very slow motion, projecting through the incompletable and inexhaustive nature of tuned intervals.

Many people who have worked with Charles can attest that things often proceed indirectly, slowly and at times insensibly, with seemingly curious tangents and asides; but when looking back over the months and years, one realizes something different and entirely unexpected has been constructed: a larger framework, and one that is surprisingly productive.

1   Alexander Grothendieck, quoted in Colin McLarty, "The Rising Sea: Grothendieck on Simplicity and Generality," in Jeremy J. Gray and Karen Hunger Parshall, eds., *Episodes in the History of Modern Algebra (1800–1950)* (Providence, RI: American Mathematical Society, 2001).

2   Grothendieck, quoted in McLarty, "The Rising Sea."

3   This essay was initially published as liner notes to *Charles Curtis: Performances and Recordings 1998–2018*, a three-CD box set released by Saltern in 2020.

4   L. E. J. Brouwer, "Consciousness, Philosophy and Mathematics," in *Collected Works* (New York: American Elsevier, 1975), vol. 1, 474.

Minimalism
and Foundations
93

Henry Flynt
Illusion-Ratios (6/19/61)

An "element" is the facing page (with the figure on it) so long as the apparent, perceived ratio of the length of the vertical line to that of the horizontal line (the element's "associated ratio") does not change.

A "selection sequence" is a sequence of elements of which the first is the one having the greatest associated ratio, and each of the others has the associated ratio next smaller than that of the preceding one. (To decrease the ratio, come to see the vertical line as shorter relative to the horizontal line, one might try measuring the lines with a ruler to convince oneself that the vertical one is not longer than the other, and then trying to see the lines as equal in length; constructing similar figures with a variety of real [measured] ratios and practicing judging these ratios; and so forth.) [Observe that the order of elements in a selection sequence may not be the order in which one sees them.]

First published as *Concept Art Version of Mathematics System 3/26/61* in the essay "Concept Art" in *An Anthology of Chance Operations* (1963), edited by La Monte Young.

Courtesy the artist

In *The Continuum*, Hermann Weyl notes:

> The states of affairs with which mathematics deals are, except
> from the very simplest ones, so complicated that it is practically
> impossible to bring them into full givenness in consciousness and, in
> this way, to grasp them completely.[1]

While a gap between the conceptual world of mathematics and its "given-ness in consciousness" is often assumed, from time to time this distance has proven a source of mathematical interest. For instance, Weyl and Brouwer, unsettled by the disparity between the classical line and intuition, sought out mathematical machinery to model the experienced continuum. More gener-ally, Brouwer's intuitionism of the 1910s and 1920s introduced an entire mathematical framework that was both time- and subject-dependent.

Although not widely adopted, Brouwer's reorientation of mathematics to include an idealized subject and his critique of formalism have intriguing, and in some cases explicit, connections to music and art of the 1960s and 1970s. In particular, the time and subject dependent form of minimalist composition developed by La Monte Young was later reinterpreted in light of such foundational concerns. This paper discusses the origins of Young's distinctive style and considers its foundational turn in works by two artists of his milieu, Henry Flynt and Catherine Christer Hennix. Flynt's concept art introduces time and subject dependent proof systems as a critique of formalism in art and mathematics, where Hennix's minimalist compositions of the 1970s theorize compositional practices in Young's music in terms of Brouwer's construction of intuitionistic sets.

I

In the summer of 1958, Young composed *Trio for Strings* on the Royce Hall organ at the University of California, Los Angeles. Notable for its focus on harmony to the exclusion of melodic considerations, the over-fifty-minute composition is made entirely of sustained harmonic groupings and silences.

Often regarded as the first piece of Minimalism, *Trio* is also a strict twelve-tone composition.[2] While the twelve-tone technique is perhaps best under-stood as method of successive variation, the key concept Young draws upon from this process of transformation is *invariance*. Hence the twelve notes in the row are subdivided into four pitch sets, and transformations are selected in a way so that as few harmonic groupings as possible occur.[3] In addition, each of the four pitch sets are subsets of the same four-note chord: a fifth with a nested fourth, and semitone in between. The logical framework of the composition rests on a single harmonic grouping, later referred to as a "Dream Chord."

Trio achieves a focus on harmonic identity through elegant and "simple" formal means. While Minimalism is sometimes associated with such logical reductions of form, this is not the approach Young himself comes to favor. The external relations of time and "musical space" present in *Trio*—such as mirror symmetries along time axes and reciprocal relations between silences and chord groupings under row transformations—cease to appear in Young's music after this piece.[4] Even in *Trio*, one can sense Young moving toward a more subject-dependent approach to time and form. The lengthy silences, sustained harmonic groupings, and use of invariance all diminish the sense of a global musical space and an external process of transformation.

In 1960, Young moves to New York and begins writing short word pieces, many of which are published together as *Compositions 1960*.[5] In these pieces, the notion of a composition as a completed form is superseded by questions of existence, performance ritual, and extra-musical activity specified within a performance context. Several pieces suggest potentially incompletable constructions of the most basic elements of music, geometry, and arithmetic. For instance, *Composition 1960 #7* notates a perfect fifth "to be held for a long time," and *Arabic Numeral (Any Integer) to H.F.* describes a loud piano cluster to be repeated some given number of times with as little change as possible. Here the notion of invariance under transformation is reoriented explicitly within a given perceptual sphere.

Perhaps most suggestive, *Composition 1960 #10 to Bob Morris* provides the instructions "draw a straight line and follow it." While this could easily be taken as a conceptual exercise, it is reflective of Young's compositional process that the piece is not only performed but carried out in a highly constructive manner. In the initial 1961 realization of the work at both Harvard University and Yoko Ono's Chambers Street loft, a sight (in this case, a vertical string tied from the floor to the ceiling) was determined, along with a point in the vicinity of where the line should end. Every few feet a plumb-bob was aligned visually with the sight, with Young providing verbal directions about adjusting the plumb. Chalk markings were made on the floor, and later all markings were connected with a yardstick.[6] As in the process of tuning, the line is only built up over time through successive perceptual adjustment. While an elementary form is investigated, it is not treated as an external reality referred to by performance, but rather something constructed in time through the subject's perspective.

In 1962, Young encounters just intonation, a system of tuning in which musical intervals are understood in terms of whole-number ratios.[7] From this point on the audible structure of the harmonic series becomes a central principle of organization in Young's music. The addition of tuning suggests an important refinement in Young's approach: not only are forms of music unfinished, but *the elements themselves* are incomplete. Comparing tuning to the astronomical observation of planets in orbit, Young notes:

> Tuning is a function of time. Since tuning an interval establishes the relationship of two frequencies in time, the degree of precision is proportional to the duration of the analysis, i.e. to the duration of tuning. Therefore, it is necessary to sustain the intervals for longer periods if higher standards of precision are to be achieved.[8]

Young goes on to argue that the accuracy of a tuned interval corresponds to the observed number of cycles of its periodic composite waveform.[9] The longer an interval is observed, the more developed it becomes. Intervals are not treated as completed points in "musical space," but rather subject-dependent constructions, developing in time and essentially incompletable. Viewed in this light, the sustained harmonic groupings present in *Trio* could be seen as further elaborations of intervals, rather than suspensions of preexisting forms. This notion of elements and forms developing in time is broadly applicable to Young's compositional process.

Reflecting this idea, the basic harmonic material of *Trio* is continually reexamined in Young's music.[10] In *The Four Dreams of China* (1962), different voicings of the Dream Chord are sustained in the manner of *Composition 1960 #7*. In 1984's *The Melodic Version* of *The Second Dream of The High-Tension Line Stepdown Transformer* from *The Four Dreams of China* (and further elaborations of *The Second Dream*), a Dream Chord and its subsets are again sustained, but are now represented by the ratio 18:17:16:12 and given in a form that is developing in time. At each moment, the performer may choose to hold their current note in the Dream Chord, pause, or possibly move to another note, with a set of rules determining the available choices at each moment given through the configurations occurring up to present. The piece is like a branching tree; any individual performance is but a single path through a much larger compositional framework. As we shall see in the final section, this method of composition, in place by the time of Young's mid-1960s pieces for his performance group Theatre of Eternal Music, resembles Brouwer's notion of a choice sequence. In addition, Young's notion of tuning as a function of time bears a likeness to Brouwer's intuitive continuum, where the elements are not completed atomistic points but unfinished *sequences* of observation.

II

In the fall of 1960, Young meets the twenty-year-old Harvard mathematics student Henry Flynt (the "H.F." in Young's *Arabic Numeral [Any Integer] to H.F.*). Inspired by Young's *Compositions 1960*, Flynt himself begins to write word pieces, which evolve into his concept art of 1961.[11] In these and later pieces, Flynt interprets the time- and subject-dependent constructions present in Young's music in terms of the foundations of mathematics.[12] Rudolf Carnap declared, "In logic, there are no morals. Everyone is at liberty to build up his own logic, i.e., his own form of language, as he wishes." Proceeding from this, Flynt proposes new logical systems based on colored pencil "action drawings," electronic music scores, and perceptual states.[13] Proof systems, including axioms and transformation rules, are specified, but the theorems themselves are purely aesthetic and devoid of traditional knowledge claims.

For instance, in *Concept Art Version of Mathematics System 3/26/61* (later titled *Illusion-Ratios*), an "element" of the system is defined to be a fixed perceived length-to-width ratio of the logical symbol $\perp$. Flynt calls this perceptual state an "associated ratio."[14] A "selection sequence" is specified as "a sequence of elements of which the first is the one having the greatest associated ratio, and each of the others has the associated ratio next smaller than that of the preceding one."[15] A theorem is a decreasing order of all associated ratios smaller than the initial perceived state.[16] Traditional aesthetic values associated with proofs, such as simplicity, economy of means, or novelty of conclusion, are replaced by the experience of the proof act itself. Indeed, in the logical framework of *Illusion-Ratios* there is a single theorem with a unique proof, assuming it can be constructed.

Like Young's *Arabic Numeral*, *Illusion-Ratios* requires retentions of memory of a purely experiential variety; however, one must now reconfigure these perceptual experiences, possibly out of the temporal ordering in which they initially appeared. Instead of suggesting a continued construction of the basic elements of a system through the subject's perspective, Flynt's logical framework further entails recognition of this process of subjectivity. He writes:

> The culture of tuning, which Young transmitted by example to his acolytes, let conscious discernment of an external process define the phenomenon. The next step is to seek the laws of conscious discernment, or recognition, of the process.[17]

The proof procedure makes formal derivation, sometimes offered as a reliable substitute for intuition, dependent on the subject's discernment of experience.

While logic typically concerns itself with the interactions between formal derivation and models of a theory, with much thought occurring on the level of models, *Illusion-Ratios* is presented purely syntactically, with no independent notion of a model. The subject's attention is focused on the experience of the symbols themselves, apart from any external reference or intended meaning. This subjective process plays an important role in the development of concept art as a whole.

In Flynt's *Derivation* (1987), the logical framework of *Illusion-Ratios* is reformulated in terms of Necker Cubes, two-dimensional line drawings that can be seen as having two distinct orientations.[18] In *Necker-Cube Stroke-Numeral* (1987), David Hilbert's view of the natural numbers as "number-signs, which are numbers and . . . objects of consideration, but otherwise have no meaning at all" is refigured in terms of a new subjective perceptual counting system, with Necker Cubes taking the place of number-signs.[19] In line with Brouwer's first act of intuitionism, separating mathematics from mathematical language, *Stroke-Numeral* criticizes Hilbert's association of formal consistency with mathematical existence.

More generally, Flynt aligns his new perceptual logical systems with a criticism of mathematical formalism that is informed by the work of Young and John Cage—in his parlance, "applying new music to metamathematics."[20] He writes:

> What did Hilbert and Carnap do? Implicitly, they cut the content out of mathematics, leaving only a formal shell. Cage anyone?[21]

However, while Hilbert's formalism sought to ensure mathematical existence by abstracting mathematics to a formal language and detecting mathematical patterns in this language, concept art ties formal syntax directly to experience, in effect blocking this process of abstraction. Flynt connects mathematical formalism with "structure art" (such as total serialism) and process art, which proceed syntactically but introduce knowledge claims at the level of metalanguage.[22] In contrast to "structure art," Flynt argues that Young's word pieces "concern the metasyntax of music. [Not *using* the rules that define music, but twisting the rules]."[23] Similarly, Cage's use of chance procedures, and letting the subject's attention define the composition, calls into question the notion of a composition as an external set of relations. In a similar way, concept art navigates "unexplored regions of formalist mathematics"—or, more poignantly, performs "'a Cage' on Hilbert and Carnap" by relativizing formal systems to the subject's attention, drawing focus upon the mind's presentational powers, apart from any external framework of relations.[24]

III

In 1969, Young meets the twenty-one-year-old composer Catherine Christer Hennix. In the same year, he commissions her to realize one of his *Drift Studies* (1966–) at the EMS studio in Stockholm.[25] Shortly thereafter, Hennix sets out writing computer music for rationally tuned sine tones. Her method of composition is closely modeled after Young's mid-1960s compositions, but she theorizes her approach in terms of Brouwer's second act of intutionism, the construction of intuitionistic sets.

The basic tool Brouwer uses for constructing intuitionistic sets are choice sequences. Choice sequences can be understood as sequences of mathematical objects, each element of which is selected by a creating subject and of which each choice may depend on all previous choices.[26] Some sequences may follow preordained rules (lawlike sequences), while others are generated quite freely by the subject (lawless sequences). For instance, the continuum can be constructed by successively choosing nested closed intervals of the form

$$\lambda_n = \left[\frac{a}{2^n}, \ \frac{a+2}{2^n} \right], \ a \in \mathbb{Z}, \ n \in \mathbb{N}$$

Here the real numbers are not given as completed atomistic *points*, but as *sequences* developing in time, depending on an idealized mathematician's attention. One can recognize in Young's compositional style a more immediately given and perceptual version of Brouwer's constructions. Indeed, choice sequences provide an interesting framework for understanding pieces such as *The Melodic Version* of the *Second Dream*, which are generated sequentially in time through the performer's continued attention to and memory of the occurring configurations. Hesseling's description of intuitionistic sets, which "do not collect mathematical objects that may or may not have been created before, but instead give a common mode of generation for its elements," provides a surprisingly apt description of Young's compositional framework from the mid-1960s on.[27] Pieces are no longer notated or fixed in advance, but continually evolve through a given harmonic framework. A composition is not treated as a set of completed external relations, but rather as a mode of generation, always developing in time through the performer's attention. Furthermore, Young's notion of tuning as a function of time could naturally be viewed in terms of Brouwer's construction of the intuitive continuum. Intervals are not treated as relations of completed points in musical space, but rather as unfinished sequences of observation subject to further refinement. Hennix appears to have sensed such implicit connections.

In 1970, Hennix introduces her algorithmic *Infinitary Compositions*. Like Young's compositions from the mid-1960s, these pieces specify *"evolving frames of musical structures, rather than trying to obtain completeness."*[28] However, instead of the compositions developing through the subject's perspective, Hennix proposes that they may be computer-generated. As Brouwer's Creating Subject was an idealized mathematician with perfect memory and indefinite attention, Hennix views the computer as an idealized creating performer where "there are no obstacles for proceeding with infinitely long spreads of musical events, locked together by some appropriate algorithm that recursively generates each new step on the basis of the preceding ones."[29] Here, Hennix theorizes Young's generative approach explicitly in terms of intuitionism.

Hennix's general proposal was never implemented, in part due to the technological limitations of the time. However, one of her initial infinitary compositions, *The N-Times Repeated Constant Event* (also referred to as $\Box^N$), was realized as part of the installation *Brouwer's Lattice* at Moderna Museet in Stockholm in 1976.[30]

Following Young's sine tone compositions of the late 1960s (and referencing *Arabic Numeral [Any Integer] to H.F.*), the constant event is understood to be one complete cycle of a composite waveform of three rationally tuned sine tones. While in *Arabic Numeral* integers are linked to "repetition . . . in the form: 'thing in time and thing again,'" through direct experiential means, Hennix follows Brouwer in articulating a more idealized and primordial account of this procedure.[31] Brouwer writes:

> mathematics is a languageless activity of the mind having its origin
> in the basic phenomenon of the perception of a *move of time*, which
> is the falling apart of a life moment into two distinct things . . . If
> the two-ity thus born is divested of all quality, there remains the
> common substratum of all two-ities, the mental creation of the
> *empty two-ity*. This empty two-ity and the two unities of which it is
> composed, constitute the *basic mathematical systems*.[32]

From this intuition of time passing, one can generate each natural number, infinitely proceeding sequences of numbers, and even infinitely proceeding sequences of mathematical systems previously acquired.

Similarly, Hennix theorizes the moment the subject comes to intuit the fundamental process of a waveform repeating in time

> corresponds to a point in her life-world where a moment of life
> falls apart with one part retained as an image and stored by memory
> while the other part is retained as a continuum of new perceptions.[33]

In this way, Brouwer's empty two-ity is linked to the experience of waveforms divested of familiar qualities (sine tones have no harmonics), and Brouwer's primordial intuition of time passing is linked to the notion of tuning.

This analogy is pushed further in *Brouwer's Lattice* taken as a whole. Using Brouwer's language of a spread, Hennix envisions a mapping between "just intonation intervals (shruti-s) and intuitionistic *mathematical entities*, both concurrently constructed by the (intutionisitic) *Creating Subject* following an intuition of time evolutions."[34] Reflective of Young's interest in tuning through auditory detection of (sometimes remote) partials over a fixed fundamental, Hennix suggests this process as a continued labeling procedure between the set of all harmonics detected in a complicated acoustical event (such as the tambura drone), and the set of all finitely branching binary trees.[35]

It is worth noting that while Hennix follows Brouwer in introducing an idealized Creating Subject into her compositional framework, she later puts forward a generalized version of Brouwer's theory. While the computer is linked to an idealized creating performer in a fairly direct way in her 1976 thinking about the *Infinitary Compositions*, in later writings the Creating Subject is theorized much more broadly. For instance, in "Revisiting *Brouwer's Lattice* Thirty Years Later," initial segments of $\square^N$ are specified as "subjective choice sequences," and "endlessly proceeding compositions correspond[ing] to *subjective mathematical assertions* by the Creative Subject about the length of ordinal numbers."[36] Indeed, Hennix later emphasizes the freedom of the Creating Subject to create nonmathematical entities, and even the rules they choose to operate under. In line with Flynt's introduction of the subject into Hilbert's formalism, Hennix extends Brouwer's intuitionism to include additional aesthetic considerations.

Although extending the framework of art to include things that were at one time viewed as non-artistic is now common practice, this process is less standard in mathematics. Indeed, Brouwer's introduction of the subject into the framework of mathematics provides a unique example of this activity. Taken from an aesthetic standpoint, this process fits broadly in line with artistic developments of the twentieth century, from Duchamp extending through Cage and Young. Young's notion of musical intervals developing in time, and his concept of a composition as a mode of generation rather than a completed entity, each resemble more immediately perceptual versions of Brouwer's subject-dependent constructions. While this connection likely reflects a mutual interest in time as a basic compositional material (Young has frequently remarked that "time is my medium"), further connections could also be made. Flynt's subjective proof theory and Hennix's intuitionistic compositions provide examples of this and offer unique instances of mathematical frameworks approached from the perspective of artistic production.

1   Hermann Weyl, *The Continuum: A Critical Examination of the Foundations of Analysis* (Mineola, NY: Dover Books, 1994), 17.

2   In twelve-tone music, the underlying structural unit is an ordering of the twelve notes in the scale (the "tone row"), to which permutations like inversion, transposition, and retrograde are applied.

3   The row is subdivided into sets $\{C\sharp, E\flat, D\}$ $\{B\flat, F\sharp, F, E\}$, $\{B\flat, A\flat, A\}$, $\{G, C\}$. The overall form is $P_0 \to I_9 \to RI_9 \to I_4 \to RI_4 \to P_0 \to Coda$. I_9 and RI_9 are the only row operations that preserve two blocks ($\{C\sharp, E\flat, D\}$ and $\{B\flat, A\flat, A\}$) from the initial partition, and I_4, RI_4 preserve a common block to I_9 and RI_9 ($\{B\flat, A\flat, A\}$). This method of dividing the row into pitch sets and looking for invariance under row operations is characteristic of late Webern, though it was never employed in such a logically reductive manner.

4   Schoenberg, who developed the twelve-tone method in the 1920s, sometimes describes twelve-tone music as an undirected space of relations, untethered from the tonal notion of a root. In *Style and Idea*, he writes: "All that happens at any point of this musical space has more than a local effect. It functions not only in its own plane, but also in all other directions and planes, and is not without influence even at remote points," and later that "there is no absolute down, no right or left, forward or backward. Every musical configuration, every movement of tones has to be comprehended primarily as a mutual relation of sounds." See Arnold Schoenberg, *Style and Idea* (New York, NY: Philosophical Library), 109, 113.

5   These are collected in *An Anthology of Chance Operations* (Bronx, NY: La Monte Young & Jackson Mac Low, 1963), a classic publication of the early 1960s New York avant-garde edited by Young.

6   The Harvard performance was organized by Henry Flynt and carried out by Young and Robert Morris. See Charles Curtis, "The Position of the Observer: Regarding the *Compositions 1960*," *± 1961: Founding the Expanded Arts* (Madrid: Museum Nacional Centro de Arte Reina Sofia, 2013), 89.

7   For instance, an octave is assigned the frequency ratio 2:1, a fifth 3:2, a fourth 4:3. Less familiar intervals, such as 28:27, 49:48, and 64:63, which are normally heard only as overtones of a fundamental, also play an important role in Young's music.

8   La Monte Young and Marian Zazeela, *Selected Writings* (Munich: Heiner Friedrich, 1969), 7.

9   Young draws a number of interesting conclusions from this view, for instance the impossibility of tuning an equal-tempered tritone, whose frequency ratio is $\sqrt{2}:1$.

Notes
Minimalism and Foundations

10   In fact, there have been five different versions of *Trio*: the 1958 version; three versions with different configurations (1984, 2001, 2005); and a three-hour just intonation version in 2015, based on Young's original sketches for the piece.

11   Although distinct from Conceptual Art of the later 1960s, it is interesting to note Flynt's connection to the genre. Flynt is part of the artistic circle of Robert Morris and Walter de Maria at this time; each contributed to *An Anthology* (Flynt's contribution is the 1961 essay "Concept Art [Provisional Version]"). Flynt notes "there was a milieu which may have consisted only of Young, Morris, myself, and one or two others, which was never chronicled in art history." See "The Crystallization of Concept Art" in 1961, 2.

12   For instance, Flynt's *Each Point on This Line Is a Composition* (1961) appears to be a specifically foundational interpretation of Young's *Composition 1960 #9*, in which a line is printed on a notecard.

13   Rudolph Carnap, *The Logical Syntax of Language*, trans. Amethe Smeaton (Countess von Zeppelin) (New York, NY: Routledge, 2007).

14   Henry Flynt, "Concept Art Version of Mathematics System 3/26/61," in *An Anthology*, ed. La Monte Young (Bronx, NY: La Monte Young and Jackson Mac Low), 1963, 28.

15   Flynt, "Concept Art," 28.

16   Flynt later expresses this system in more familiar logical notation, stipulating "an associated ratio is a sentence," an "axiom is the first sentence one sees," and "sentence A implies sentence B if the associated ratio of B is the next smallest ratio of all sentences you see." See Henry Flynt, "Mathematics, Tokenetics, and Uncanny Calculi: 1961 Concept Art in Retrospect," last modified 1996, http://www.henryflynt.org/meta_tech/token.html.

17   Henry Flynt, "The Electric Harpsichord," in Catherine Christer Hennix, *Pöesy Matters and Other Matters*, vol. 2 (Brooklyn, NY: Blank Forms Editions, 2019), 164.

18   Flynt believes this new framework simplifies issues surrounding the continuity of perception in *Illusion-Ratios*, and subsequently the cardinality of its language.

19   Translation in Dennis Hesseling, *Gnomes in the Fog: The Reception of Brouwer's Intuitionism in the 1920s* (Basel: Birkhäuser, 2003), 140.

20   Henry Flynt, "The Crystallization of Concept Art in 1961," last modified 1994, http://www.henryflynt.org/meta_tech/crystal.html.

21   Henry Flynt, *2011 Concept Art 50 Years* (Berlin: Grimmuseum, 2011), 12.

22 The then-current offshoot of twelve-tone music, where rhythm, duration, timbre, etc., are acted on by permutations.

23 Flynt, "The Crystallization of Concept Art in 1961."

24 Flynt, "Concept Art," 28; Flynt, *2011 Concept Art 50 Years*, 12.

25 In these pieces, rationally tuned sine tones gradually go in and out of phase. Although Young refers to "tuning as a function of time" as one of his key theoretical constructs, it is interesting to note that this philosophy may have stemmed in part from the technological limitations of the time. In works such as *Dream House* (1969), Young envisions sustaining tuned intervals for weeks or longer by electronic means. However, the realization of such works proved difficult due to the instability of commercially available oscillators of the time. EMS had recently purchased phase-locked oscillators, which Young was interested in testing.

26 More generally, choice sequences can be understood in terms of spreads. A spread consists of a *spread law* Γ_M, which is a lawlike characteristic function on $\mathbb{N}^{<\mathbb{N}}$, and a *complementary law* Λ_M, which assigns a mathematical object to each finite sequence $\langle a_1, a_2, \ldots, a_n \rangle$ such that $\Lambda_M (\langle a_1, a_2, \ldots, a_n \rangle) = 1$. See Hesseling, *Gnomes in the Fog*, 65.

27 Hesseling, *Gnomes in the Fog*, 66.

28 Catherine Christer Hennix, "A Brief Presentation of *Brouwer's Lattice* and *The Deontic Miracle*," interview with Rita Knox, in *Pöesy Matters and Other Matters*, 52.

29 Hennix, "A Brief Presentation," 16. See note 26 for a discussion of Brouwer's notion of a spread.

30 Although the title is suggestive of a finite process, Hennix refers to the piece as infinitary and envisions it as composed of three infinitely sustained sine tones. See Hennix, "Revisiting *Brouwer's Lattice* Thirty Years Later," in *Pöesy Matters and Other Matters*, 259.

31 L. E. J. Brouwer, "On the Foundations of Mathematics," in *Collected Works 1: Philosophy and Foundations of Mathematics*, ed. A. Heyting (Amsterdam: North-Holland, 1975), 53.

32 L. E. J. Brouwer, "Points and Spaces," in *Collected Works 1: Philosophy and Foundations of Mathematics*, 523.

33 Catherine Christer Hennix, "☐$_\kappa$ Excerpt from Notes on the Composite Sine-Wave Drone Over Which the Electric Harpischord is Performed," in *Pöesy Matters and Other Matters*, 26.

34 Hennix, "Hilbert Space Shruti Box (of the [Quantum] Harmonic Oscillator)," in *Pöesy Matters and Other Matters*, 280.

35 For instance, a finite sequence like (1,0,0,1,0) would indicate which harmonics were detected as present or

absent, based on some enumeration of all harmonics of the fundamental. Hennix has more recently remarked that she associates the intuitive continuum with the spectrum of the tambura, the basic acoustical reference for Young's music since the 1970s.

36   Hennix, "Revisiting *Brouwer's Lattice*," 259.

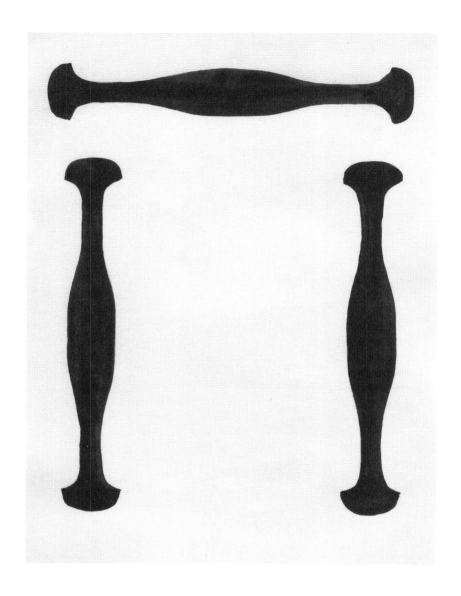

Ricky Swallow
Shaker Pegs (Arch), 2015
Gouache on paper
Courtesy of David Kordansky Gallery

Order Three Normalizers of 2-Groups

111

Abstract

This paper examines order three elements of finite groups that normalize no nontrivial 2-subgroup. The motivation for finding such elements arises out of a problem in modular representation theory. A complete classification of order three elements with this property is determined for the almost simple groups, answering a question of G. Robinson. On the basis of this result, necessary conditions are determined for the existence of such elements in a large class of finite groups.

1   Introduction

Problem 19 on Brauer's [2] well-known list of questions in representation theory asks to describe the number of blocks of defect zero of a finite group G in terms of group-theoretic invariants. Robinson [9] provides a solution to this question, however in many cases the proposed invariants are difficult to determine.

More recently, a lower bounds on 2-blocks of defect zero was obtained in terms of a separate group-theoretic property. In [10], it is shown that the number of 2-blocks of defect zero of a finite group G is at least as great as the number of conjugacy classes of elements of order three which normalize no nontrivial 2-subgroup of G.

It is not difficult to construct groups containing order three elements with this property. For instance if $x \in G$ is order three, and U is a maximal x-invariant 2-subgroup of G, then xU will normalize no nontrivial 2-subgroup of $N_G(U)/U$, and hence $N_G(U)/U$ will have a 2-block of defect zero. However, it is less clear when such elements might arise in familiar contexts. In order to better understand this situation, Robinson asks which almost simple groups contain elements of order three normalizing no nontrivial 2-subgroup. The small list of examples is described below.

Theorem 1.   Let G be a finite almost simple group, and $x \in G$ be an element of order three. Then x normalizes a nontrivial 2-subgroup of G, unless G and x occur in the following list

(i) $PSL_2(2^a)$, a odd, x in the unique class of order three elements;
(ii) $PGL_3(2^a)$, a even, x in an irreducible torus;
(iii) $PGU_3(2^a)$, a odd, x in an irreducible torus;
(iv) ${}^2G_2(q^2)$, $q^2 = 3^{2f+1}$, x in the class $(\tilde{A}_1)_3$.

Making use of **Theorem 1**, it is possible to determine necessary conditions for the existence of order three elements with the stated property in a large class of finite groups. Let $O_p(G)$ be the largest normal p-subgroup of G. Recall that a component L of G is a quasisimple subnormal subgroup. The following general restriction is established in the final section.

Theorem 2.   Assume G is a finite group with components $L_1, ..., L_n$. Furthermore, assume that $O_3(G) = 1$. Let $x \in G$ be an order three element normalizing no nontrivial 2-subgroup of G. Then

(i)   if N is any normal subgroup of G with order prime to three, then N has order prime to six;
(ii) x normalizes every component L_i of G;

(iii)   The image of $\langle L_i, x \rangle$ in $Aut(L_i)$ appears on the list of exceptions in **Theorem 1**, for every component L_i of G.

Much of the work in proving **Theorem 1** goes into determining which order three elements $x \in G$ normalize a group of order two. This is equivalent to asking whether x centralizes an involution in G. **Theorem 3** classifies the order three elements of almost simple groups that centralize no involution. Note if G is a finite classical group in characteristic $p = 3$, and S is the corresponding simple group, then $Z(G)$ has order prime to three. Hence any order three element $x \in S$ can be lifted to a unipotent element of order three $\hat{x} \in G$.

Theorem 3.   Assume $G = \langle S, x \rangle$ is a finite almost simple group, with $x \in G$ an element of order three. Then x centralizes an involution in G, unless G and x occur in **Table 1**.

Table 1

G	x
$PSL_2(2^a)$, $a \geq 1$	in the unique class of order three elements
$PSL_2(q)$, $q \equiv 5 \pmod{12}$	in the unique class of order three elements
$PSL_2(q)$, $q \equiv 7 \pmod{12}$	in the unique class of order three elements
$PSL_3(2^a)$, $a \geq 1$	in a split torus for a even;
	in a partially split torus for a odd
$PSU_3(2^a)$, $a \geq 1$	in a split torus for a odd;
	in a partially split torus for a even
$PGL_3(q)$, $q \equiv 1 \pmod{3}$	in an irreducible torus
$PGU_3(q)$, $q \equiv -1 \pmod{3}$	in an irreducible torus
$PSL_2(3^a)$, $a \geq 1$	$\hat{x}$ has Jordan form J_2
$PSL_3(3^a)$, $a \geq 1$	$\hat{x}$ has Jordan form J_3
$PSU_3(3^a)$, $a \geq 1$	$\hat{x}$ has Jordan form J_3
$PSL_4(3^a)$, a odd	$\hat{x}$ has Jordan form $J_3 \oplus J_1$
$PSU_4(3^a)$, $a \geq 1$	$\hat{x}$ has Jordan form $J_3 \oplus J_1$
Alt_5	in class (1 2 3)
Alt_6	in class (1 2 3), or (1 2 3)(4 5 6)
Alt_7	in class (1 2 3)(4 5 6)
Alt_n, $n = 9, 10,$	in class (1 2 3)(4 5 6)(7 8 9)
the Janko group J_3	in class 3B from the J_3 table in [3]
$G_2(q)$, $q = 3^f$	in class $(\tilde{A}_1)_3$ from **Table 22.2.6** of [6]
$^2G_2(q^2)$, $q^2 = 3^{2f+1}$	in class $(\tilde{A}_1)_3$ from **Table 22.2.7** of [6]

Let us briefly describe the order in which the main theorems are proved. The paper begins with a short section on centralizers of order three elements in the alternating and sporadic groups. The next and most lengthy section determines all odd order centralizers of order three elements in the finite simple groups of Lie type. Results in this section are proved by working in the relevant algebraic groups, and reducing to the finite simple group case. The arguments differ substantially in characteristics $p = 3$, and $p \neq 3$. Next, there is a section considering outer automorphisms of order three. Taken together, these first three sections establish **Theorem 3**. In the final section, **Theorem 1** is proved by checking whether the list of exceptions found in **Table 1** of **Theorem 3** normalize some larger 2-subgroup. **Theorem 1** is then used in order to establish **Theorem 2**.

2   Alternating and Sporadic Groups

In this brief section we determine all order three elements with odd order centralizers in the alternating and sporadic groups. The following observation restricts our search for order three elements in Alt_n that centralize no involution.

Lemma 1.   Assume $x \in \mathrm{Alt}_n$ is order three. Then x centralizes an involution for $n > 12$.

> **Proof.**   If x has at least four fixed points, then x will central-ize a product of two disjoint 2-cycles, and hence an involution in Alt_n. Next, an element of the form $(123)(456)(789)(10\ 11\ 12)$ is central-ized by the involution $(14)(25)(36)(7\ 10)(8\ 11)(9\ 12)$. Hence any order three element containing four or more three cycles will centralize an involution. For $n > 12$, all elements of order three in Alt_n contain either four fixed points or four disjoint three cycles.   □

By an easy inspection, we are led to the following.

Lemma 2.   Let $S = \mathrm{Alt}_n$ and assume $x \in S$ is order three. Then $C_S(x)$ is even order, unless $n = 5, 6, 7, 9$ or 10.

The ATLAS [3] lists the centralizer orders of order three elements in the sporadic groups. This yields the following conclusion.

Lemma 3.   Assume S is a sporadic group, and $x \in S$ is order three. Then $C_S(x)$ is even order, unless $S = J_3$ and x is in the conjugacy class $3B$.

3   Groups of Lie Type

In this section we determine the order three elements with odd order centralizers in the finite simple groups of Lie type. As noted earlier, the arguments make frequent use of the relevant algebraic groups. Unless otherwise stated, let X be a simple algebraic group of simply connected type, $X^F = G$ be the fixed points of a Steinberg endomorphism $F : X \rightarrow X$, and $S = G/Z(G)$.

Then S is a finite simple group, unless G appears in the following list (see [8], Theorem 24.17):

$$SL_2(2), SL_2(3), SU_3(2), Sz(2), Sp_4(2), G_2(2), {}^2G_2(3) \text{ or } {}^2F_4(2)$$

In the first four cases listed above, the group G is solvable. These cases are not relevant to **Theorem 3**. In the latter four cases, $G/Z(G)$ is not simple but the derived subgroup G' is. As $Sp_4(2)' \cong \mathrm{Alt}_6$, $G_2(2)' \cong PSU_3(3)$ and ${}^2G_2(3)' \cong PSL_2(8)$, these groups are considered elsewhere. From the ATLAS [3], we find ${}^2F_4(2)'$ has a unique class of order three elements, and elements in this class have even order centralizers. Hence in what follows it suffices to limit our attention to cases where $S = G/Z(G)$ is simple.

Using some representation, view G as a matrix group over the field $k = \mathbb{F}_q$, where $q = p^n$, and p is the underlying characteristic of G. Recall $x \in G$ is *semisimple* if it is diagonalizable over the algebraic closure of k, *unipotent* if all its eigenvalues are 1, and *regular* if the dimension of its centralizer in X is as small as possible. Note when $x \in G$ is order three and $p = 3$, $(x - I)^3 = x^3 - I = 0$, with I the identity matrix. So x is unipotent. Similarly when $p \neq 3$, $3 \mid q^2 - 1$, so an order three element is diagonalizable over $\mathbb{F}_{q^2}$, and hence semisimple. It will be convenient to treat these two cases separately.

Next, let $\pi : G \to G/Z(G)$ be the natural projection map, and $\hat{x}$ be an element in the preimage of $x \in G/Z(G)$. We will say $x \in S$ is semisimple, unipotent, or regular, if some element $\hat{x} \in G$ is. The following lemma shows that the question of whether an order three element $x \in S$ centralizes an involution can be lifted to the question of whether an odd order element $\hat{x} \in G$ commutes with a non-central 2-power element.

Lemma 4.   Let G be a finite group, $x \in G$ be an element of odd order, and w be a 2-power element of G. Then x commutes with w modulo $Z(G)$ if and only if x commutes with w in G.

> **Proof.**   Assume $x \in G$ has odd order r, and $w \in G$ has order $s = 2^k$, for some $k \geq 1$. If $[w, x] \in Z(G)$, then $[w, x^r] = [w, x]^r = 1$, and $[w^s, x] = [w, x]^s = 1$. Since r and s are relatively prime, this implies $[w, x] = 1$, and $w \in C_G(x)$. Conversely, if $w \in C_G(x)$, then $[w, x] = 1 \in Z(G)$.   □

In particular, when $Z(G)$ has order prime to three, showing S contains no order three elements with odd order centralizers is equivalent to showing all order three elements in G commute with a non-central 2-power element. This fact will be useful in treating the unipotent classes below.

3.1   Semisimple Classes

Assume S is a finite simple group of Lie type in characteristic $p \neq 3$, and $x \in S$ is an element of order three. The following lemma shows $C_S(x)$ is even order, except in a few small rank cases.

Lemma 5. Let X be a simply connected simple algebraic group, $X^F = G$ be the fixed points of a Steinberg endomorphism $F : X \to X$, and $S = G/Z(G)$ be a finite simple group. If $x \in S$ is semisimple but not regular semisimple, then $C_S(x)$ is even order.

> **Proof.** Consider $\hat{x} \in X^F$. Since X is simply connected and $\hat{x}$ is semisimple but not regular semisimple, $C_X(\hat{x})$ is a connected, reductive algebraic group strictly containing a torus (see [5], Sections 2.2 and 2.11). Hence $C_X(\hat{x})$ contains a semisimple algebraic subgroup Y. The Weyl group of Y^F contains elements of even order, which ensures $C_S(x)$ contains an involution. □

Note **Lemma 5** holds more generally in cases where $G/Z(G)$ is not solvable, but $G'/Z(G')$ is simple. In this situation, if $\hat{x} \in G$ is semisimple but not regular semisimple, $C_X(\hat{x})$ contains a semisimple algebraic subgroup Y and the Weyl group of Y^F contains a Sylow 2-subgroup whose order does not divide $[G : G']$. Again $C_{G'/Z(G')}(x)$ has even order.

Corollary 1. Let $S = G/Z(G)$ be a finite simple group, with X and $X^F = G$ as above. Assume $x \in S$ is a semisimple order three element. Then $C_S(x)$ is even order, unless $S = PSL_2(q), PSL_3(q)$ or $PSU_3(q)$.

> **Proof.** To begin, assume $X = SL_n(k)$ or $Sp_n(k)$, and V is the natural module for X. If $x \in S$ is semisimple of order three, then $\hat{x} \in G$ has at most three distinct eigenvalues on V. As regular semisimple elements in X have no repeated eigenvalues on the natural module, X contains no regular semisimple order three elements for $n \geq 4$.
>
> Next let $X = Spin_n(k)$, $Y = SO_n(k)$, and $\rho : X \to Y$ be the standard covering map (so that ρ is a double cover in characteristic $p \neq 2$). Note if $y \in Y$ is not regular, then any preimage $\hat{y} \in X$ will share this property. Let $Y^F = H$ be the fixed points of a Steinberg endomorphism $F : Y \to Y$, and $R = H'/Z(H')$ be the corresponding finite simple group. A regular semisimple element in Y has no repeated eigenvalues on the natural module other than ± 1, and at most one eigenspace of dimension two corresponding to an eigenvalue ± 1. Hence X and Y contain no regular semisimple elements of order three for $n \geq 5$. In addition, $Spin_4(k)$ is not a simple algebraic group ($Spin_4(k)/Z(Spin_4(k)) \cong PSL_2(k) \times PSL_2(k)$), and $Spin_3(k) \cong Sp_2(k) \cong SL_2(k)$.

Hence if X is a simple classical algebraic group of simply connected type and $x \in S$ is a regular semisimple element of order three, then $X = SL_3(k)$ or $X = SL_2(k)$, and $S = PSL_2(q), PSL_3(q)$ or $PSU_3(q)$. From Lübeck [7], there are no regular semisimple elements of order three in the simply connected exceptional algebraic groups. The conclusion follows by applying **Lemma 5**. $\square$

We may now classify the semisimple elements of order three with odd order centralizers.

Theorem 4. Let X be a simply connected simple algebraic group, $X^F = G$ be the fixed points of a Steinberg endomorphism $F : X \rightarrow X$, and $S = G/Z(G)$ be a finite simple group. Assume $x \in S$ is a semisimple order three element. Then $C_S(x)$ is even order, unless S and x occur in the following list.

(i) $PSL_3(2^a)$, $a \geq 1$, x in a split torus for a even; x in a partially split torus for a odd

(ii) $PSU_3(2^a)$, $a \geq 1$, x in a split torus for a odd; x in a partially split torus for a even

(iii) $PSL_2(2^a)$, $a \geq 1$, x in a split torus for a even; x in an irreducible torus for a odd

(iv) $PSL_2(q)$, $q \equiv 7 \pmod{12}$, x in a split torus

(v) $PSL_2(q)$ $q \equiv 5 \pmod{12}$, x in an irreducible torus

Proof. Applying **Lemma 5** and **Corollary 1**, any order three element $x \in S$ will centralize an involution unless (a) $S = PSL_2(q)$, $PSL_3(q)$ or $PSU_3(q)$, and (b) x is regular semisimple. So assume x, S satisfy conditions (a) and (b). Since X is a simply connected algebraic group, $\hat{x}$ will live in some maximal torus Q, with $C_X(\hat{x})^F = Q$ (see [5], **Section 2.11**). Let $T = Q^F/Z(X^F)$ be the corresponding maximal torus in S. Then $C_S(x)$ is odd order if and only if T is. Hence to classify semisimple order three elements with odd order centralizers, it suffices to find all odd order tori in $PSL_2(q), PSL_3(q)$ or $PSU_3(q)$ containing regular semisimple elements of order three.

To begin, assume $S = PSL_3(q)$. Up to conjugacy, S has three maximal tori; split, partially split, and irreducible. Call these tori T_1, T_2 and T_3.

A split torus T_1 has order $\frac{(q-1)^2}{(3,q-1)}$, and hence is odd order if and only if q is even. Furthermore, T_1 contains elements of order three if and only if $q \equiv 1 \pmod 3$. So assume $q \equiv 1 \pmod 3$ is even, and $\alpha \in \mathbb{F}_q^\times$ has order three. Then some $\hat{x} \in X^F$ has eigenvalues $1, \alpha, \alpha^2$ on the natural module, and hence T_1 contains regular semisimple elements of order three. It follows that $S = PSL_3(2^a)$, a even, contains elements of order three centralizing no involution.

A partially split torus T_2 has order $\frac{q^2-1}{(3,q-1)}$, and so is odd order only if q is even. First assume q is even, and $q \equiv -1 \pmod 3$. Let α be an element of order three in $\mathbb{F}_{q^2}^\times / \mathbb{F}_q^\times$. Some $\hat{x} \in X^F$ is diagonalizable in $SL_3(q^2)$ with eigenvalues $1, \alpha, \alpha^2$ on the natural module. In this case, T_2 contains regular semisimple elements of order three. On the other hand, when $q \equiv 1 \pmod 3$, T_2 contains no regular elements of order three. So $S = PSL_3(2^a)$, a odd, contains elements of order three centralizing no involution.

An irreducible torus T_3 has order $\frac{q^2+q+1}{(3,q-1)}$. If $q \equiv -1 \pmod 3$, then $3 \nmid q^2 + q + 1$. If $q \equiv 1 \pmod 3$, then $3 \mid q^2 + q + 1$, but $9 \nmid \frac{q^2+q+1}{(3,q-1)}$. So T_3 contains no elements of order three.

Next consider $S = PSU_3(q)$. The three maximal tori in S have orders $\frac{(q+1)^2}{(3,q+1)}$, $\frac{q^2-1}{(3,q+1)}$, and $\frac{q^2-q+1}{(3,q+1)}$. Repeating the arguments given above yields that $C_S(x)$ is even order, unless $S = PSU_3(2^a)$. When a is odd, there are elements of order three in a split torus, and when a is even, there are elements of order three in a partially split torus.

Finally, consider $S = PSL_2(q)$. S contains two types of tori, split and irreducible, of orders $\frac{q-1}{(2,q-1)}$ and $\frac{q+1}{(2,q-1)}$. First assume q is even. Then both tori have odd order and there are regular semisimple elements of order three in a split torus when $q \equiv 1 \pmod 3$, and regular semisimple elements of order three in an irreducible torus when $q \equiv -1 \pmod 3$. So $PSL_2(2^a)$, $a \geq 1$, yields additional classes of exceptions. Next assume q is odd. There are regular semisimple elements of order three in a split torus when $q \equiv 7 \pmod{12}$, and in an irreducible torus when $q \equiv 5 \pmod{12}$. Furthermore in both cases these tori have odd order. This yields two final classes of exceptions. $\qquad\square$

3.2   Unipotent Classes

Now let X be a simply-connected simple algebraic group in characteristic $p = 3$, with $X^F = G$ and $S = G/Z(G)$ as above. In this situation $Z(G)$ has order prime to three, and every order three element $x \in S$ can be lifted to a unipotent element of order three $\hat{x} \in G$. Hence by **Lemma 4** to show every order three element in S centralizes an involution, it suffices to show all order three elements in G commute with a non-central 2-power element.

For $X = Spin_n(k)$ it will be convenient to switch from the simply-connected algebraic group to the isogeny type $X = SO_n(k)$. Letting $X^F = G$ be as above, we have that $G/Z(G)$ is not simple, but $S = G'/Z(G')$ is. In this situation, to show every order three element $x \in S$ centralizes an involution, it suffices to show all order three elements in G commute with a non-central 2-power element of spinor norm 1.

Assume $x \in G$ is unipotent of order three. We show $C_G(x)$ contains a 2-power element with the desired properties, except in a few small rank cases. For the classical groups, the relevant criterion is whether x has a repeated Jordan block. The following theorem will be used on several occasions.

Theorem 5. Assume $X = GL_n(k)$, $Sp_n(k)$, or $O_n(k)$ with k an algebraically closed field of characteristic $p \neq 2$. Assume $x \in X$ is a unipotent element with Jordan form $\oplus_i J_i^{r_i}$.

(i) If $X = Sp_n$, then r_i is even for each odd i; and if $X = O_n$ then r_i is even for each even i.

(ii) Let $C_X(x) = UR$, with U the unipotent radical $R_u(C_X(x))$, and R the reductive part of the group. Then,

$$R = \prod GL_{r_i}, \text{ if } X = GL_n.$$
$$R = \prod_{i \ odd} Sp_{r_i} \times \prod_{i \ even} O_{r_i}, \text{ if } X = Sp_n.$$
$$R = \prod_{i \ odd} O_{r_i} \times \prod_{i \ even} Sp_{r_i}, \text{ if } X = O_n.$$

Proof. See **Theorem 3.1** from [6]. □

Lemma 6. Assume $X = SL_n(k)$, $Sp_n(k)$, or $SO_n(k)$ is a simple algebraic group, and $x \in X$ is a unipotent order three element with a repeated Jordan block. Then $C_X(x)$ contains a simple algebraic subgroup, unless $X = Sp_4(k)$, $SO_7(k)$, or $SO_8(k)$.

Proof. Applying **Theorem 5**, if $X = SL_n(k)$ and $x \in X$ has a repeated Jordan block, then $C_X(x)$ contains a copy of $SL_2(k)$.

For $X = Sp_n(k)$, odd sized blocks come in pairs, and any odd pair ensures the existence of $Sp_2(k) \cong SL_2(k)$ in $C_X(x)$. So we may assume $x \in X$ contains only even sized Jordan blocks. Since x has order three, this further implies that x contains only Jordan blocks of size two. If x contains three blocks of size two, then $C_X(x)$ contains a copy of $O_3(k) \cong PGL_2(k)$. The only remaining case is $X = Sp_4(k)$, $x = J_2 \oplus J_2$, where $C_X(x)/R_u(C_X(x)) \cong O_2(k)$ is solvable.

If $x \in SO_n(k)$ has a repeated Jordan block of size two, or three repeated blocks of size one or three, then $C_X(x)$ contains a copy of $SL_2(k)$. As even blocks come in pairs, we may assume that only odd blocks occur. Since $SO_2(k)$ is not simple, and $Spin_6(k) \cong SL_4(k)$, the only remaining cases to check are $X = SO_7(k)$, $x = J_3 \oplus J_3 \oplus J_1$, and $X = SO_8(k)$, $x = J_3 \oplus J_3 \oplus J_1 \oplus J_1$. In these situations, $C_X(x)/R_u(C_X(x))$ is solvable. □

Lemma 7. Let $X = SL_n(k)$, $Sp_n(k)$, or $SO_n(k)$ be a simple algebraic group, with $X^F = G$ as above. Assume $x \in G$ is unipotent of order three. If x

contains a repeated Jordan block, then x commutes with a non-central 2-power element in G. Furthermore, if $X = SO_n(k)$ we may assume this non-central 2-power element has spinor norm 1.

Proof.   Assume $x \in G$ has a repeated Jordan block, and $C_X(x)$ contains a simple algebraic subgroup Y. In characteristic $p = 3$, for any such $Y \subset C_X(x)$ the order of $Y^F \subset C_G(x)$ is divisible by eight. Since $|G|/|S| \leq 4$, $C_G(x)$ must contain a non-central 2-power element (of spinor norm 1, if $X = SO_n(k)$). It then follows from **Lemma 6** that the only cases requiring attention are: $X = Sp_4(k)$, $x = J_2 \oplus J_2$; $X = SO_7(k)$, $x = J_3 \oplus J_3 \oplus J_1$; and $X = SO_8(k)$, $x = J_3 \oplus J_3 \oplus J_1 \oplus J_1$.

Let $X = Sp_4(k)$ and $x = J_2 \oplus J_2$. **Table 8.1a** of [6] lists the orders of centralizers of unipotent classes in $G = Sp_4(q)$. For $J_2 \oplus J_2$ there are two classes, with centralizer orders $2q^3(q \pm 1)$. Since $Z(G)$ has order two, and four divides the order of $C_G(x)$, x commutes with a non-central 2-power element.

Next assume $X = SO_7(k)$, $x = J_3 \oplus J_3 \oplus J_1$. **Table 8.4a** of [6] says the $SO_7(q)$ classes of x have centralizer orders $2q^6(q \pm 1)$. Since the centralizer order contains a factor of four, and S has index two in $SO_7(q)$, x commutes with a non-central 2-power element of spinor norm 1. Finally, if $X = SO_8(k)$ and $x = J_3 \oplus J_3 \oplus J_1 \oplus J_1$, Table 8.5a of [6] tells us the two $SO_8^+(q)$ classes of x have centralizers orders $2q^8(q \pm 1)^2$, and the two $SO_8^-(q)$ classes of x have centralizer orders $2q^8(q^2 - 1)$. Again since eight divides the order of the centralizer and $|SO_8^{\pm}(q)|/|S| \leq 4$, x commutes with a non-central 2-power element of spinor norm 1.   □

We are now in a position to determine the unipotent elements of order three with odd order centralizers.

Theorem 6.   Assume S is a finite simple group of Lie type in characteristic $p = 3$, and $x \in S$ has order three. Then $C_S(x)$ is even order, unless S and x occur in the following list.

(i) $PSL_2(3^a)$, $a \geq 1$, $\hat{x}$ has Jordan form J_2
(ii) $PSL_3(3^a)$, $a \geq 1$, $\hat{x}$ has Jordan form J_3
(iii) $PSU_3(3^a)$, $a \geq 1$, $\hat{x}$ has Jordan form J_3
(iv) $PSU_4(3^a)$, $a \geq 1$, $\hat{x}$ has Jordan form $J_3 \oplus J_1$
(v) $PSL_4(3^a)$, a odd, $\hat{x}$ has Jordan form $J_3 \oplus J_1$
(vi) $G_2(q)$, $q = 3^f$, x in class $(\tilde{A}_1)^3$
(vii) $^2G_2(q^2)$, $q^2 = 3^{2f+1}$, x in class $(\tilde{A}_1)^3$

Proof.   First assume X is an exceptional algebraic group. Lübeck [7] shows in characteristic $p = 3$, all classes of order three elements have Bala-Carter parameters involving root systems of type $A_1, \tilde{A}_1, A_2, \tilde{A}_2$, or $G_2(a_1)$. Tables 22.2.1- 22.2.7 in [6] list the centralizer orders in X^F of classes with such parameters. For $X \not\equiv G_2(k)$, all relevant centralizers have even order. For $X^F \cong G_2(q)$ and $X^F = {}^2G_2(q^2)$, $(\tilde{A}_1)^3$ is the only class of order three elements with an odd order centralizer.

Next assume X is a simple classical algebraic group $SL_n(k), Sp_n(k)$, or $SO_n(k)$. Let S be the corresponding finite simple group, and assume $x \in S$ has order three. By lifting $x \in S$ to a unipotent element of order three $\hat{x} \in G$ and applying **Lemma 7**, we may assume $\hat{x}$ contains no repeated Jordan blocks, and hence that $n \le 6$.

First let $X = SL_n(k)$. Note if $S = PSL_n(q)$ or $PSU_n(q), n > 2$, and $\hat{x}$ contains a J_2 block, then there exists a non-central involution $diag[1, ..., -1, -1, ..., 1]$ in G commuting with $\hat{x}$ (this element occurs in $SU_n(q)$, since $3^a + 1$ is even). Hence we may further restrict to classes in $SL_n(q)$ and $SU_n(q)$ containing no Jordan blocks of size two. This limits our search to $n \le 4$.

When $n = 2$ or 3, $\hat{x}$ must have Jordan forms J_2 or J_3, respectively. In these cases, the reductive part R of $C_G(\hat{x})$ is contained in $Z(G)$. Since the unipotent radical has odd order, $C_S(x)$ will contain no involutions. Hence $PSL_2(3^a), PSL_3(3^a), PSU_3(3^a), a \ge 1$, yield classes of exceptions.

When $n = 4$, $\hat{x}$ must have Jordan form $J_3 \oplus J_1$. In $SL_4(k)$, the reductive part of the centralizer of $\hat{x}$ will consist of diagonal matrices of the form $diag[\alpha, \alpha, \alpha, \beta]$, where $\alpha^3 \beta = 1$. Hence if y is an involution modulo the center and is contained in $C_G(\hat{x})$, we must have that $y = diag[\alpha, \alpha, \alpha, \beta]$, with $\alpha^2 = \beta^2$. Note that $\alpha^3 \beta = 1$ and $\alpha^2 = \beta^2$ imply that $\alpha^8 = 1$. If α has order one, two, or four, the above conditions further imply $\alpha = \beta$, and y is the identity. Hence $C_S(x)$ will have odd order, unless α has order eight.

When a is odd, $\mathbb{F}_{3^a}^\times$ contains no elements of order eight, so $PSL_4(3^a)$, a odd, yields another class of exceptions. When a is even, $\mathbb{F}_{3^a}^\times$ does contain an element α of order eight, and $diag[\alpha, \alpha, \alpha, \alpha^5]$ is a non-central involution commuting with $\hat{x}$.

Elements $diag[\alpha, \alpha, \alpha, \beta]$ in $SU_4(q)$ have the additional property that the orders of α and β must divide $q + 1$. Since $8 \nmid q + 1$, this implies that $PSU_4(3^a), a \ge 1$, yields a further class of exceptions.

Next consider the symplectic groups $X = Sp_4(k)$ and $Sp_6(k)$. Theorem 5 says $J_3 \oplus J_1 \notin Sp_4(k)$, and $J_3 \oplus J_2 \oplus J_1 \notin Sp_6(k)$. So elements of order three will have repeated Jordan blocks, and hence centralize an involution. Finally, there are exceptional isomorphisms allowing us to handle $X = SO_n(k)$, $n \leq 6$, in terms of the other classical groups. □

Applying **Lemmas 2** and **3**, **Theorems 4** and **6**, and the Classification of Finite Simple Groups, we have determined the order three elements in the finite simple groups which centralize no involution. To complete the proof of **Theorem 3**, it suffices to find all order three elements in $S < G \leq \operatorname{Aut}(S)$ with odd order centralizers.

4 Outer Automorphisms of Order Three

Lemma 8. Let $G = \operatorname{Sym}_n$, with $n \geq 5$. If $x \in G$ is order three, then $C_G(x)$ has even order.

Proof. If $x \in G$ has two fixed points or two three cycles, it will centralize an involution in G. For $n \geq 5$, all order three elements have this property. □

Note for $n \geq 5$, Sym_n is almost simple, and for $n \neq 6$, the outer automorphism group of Alt_n is Sym_n. When $n = 6$, the automorphism group is slightly bigger, however this case is handled elsewhere by the exceptional isomorphism $\operatorname{Alt}_6 \cong PSL_2(9)$.

Lemma 9. Let $S < G \leq \operatorname{Aut}(S)$ be almost simple, with S a sporadic group. Assume $x \in G$ is order three. Then $C_G(x)$ has even order.

Proof. From [11] we find each order three element in G has an even order centralizer. □

It only remains to check order three outer automorphisms in the groups of Lie type. For this we must consider: *(i)* diagonal automorphisms, *(ii)* field automorphisms, *(iii)* graph automorphisms, and *(iv)* field-graph automorphisms. The following lemma shows the latter three cases contain no elements of order three with odd order centralizers.

Lemma 10. Let $S < G \leq \operatorname{Aut}(S)$ be an almost simple group, with S a simple group of Lie type. Assume $x \in G$ is an order three field, graph, or field-graph automorphism. Then $C_G(x)$ has even order.

Proof. If S is a finite simple group of Lie type over the field $\mathbb{F}_{p^n}$, then the field automorphism group of S is cyclic of order n. Hence there is an almost simple group $G = \langle S, x \rangle$ containing an order three field automorphism x whenever three divides n. However, field automorphisms fix the prime field, and hence an even order subgroup $Y \subset S$. So $C_G(x)$ has even order.

Order three graph and field-graph automorphisms occur only in $D_4(q)$ (see **Section 4.7** of [4]). In characteristic $p = 3$, there are two conjugacy classes of graph automorphisms in D_4 (see **Proposition 4.9.2** in [4]). The first has fixed point group $G_2(q)$, and the second has a parabolic subgroup of $G_2(q)$ as its fixed point group. Hence both are even order for all q. In characteristic $p \neq 3$, the two conjugacy classes of graph automorphisms have fixed point groups $G_2(q)$ and $A_2(q)$ (see **Theorem 4.7.1** in [4]). These groups both contain involutions for all q. Similarly, up to diagonal automorphism, there is a single class of order three graph-field automorphisms in $D_4(q)$. The centralizer of an order three graph-field automorphism is $^3D_4(q)$, which contains involutions for all q. □

Finally, consider the order three diagonal automorphisms.

Lemma 11.   Let $S < G \leq \operatorname{Aut}(S)$ be an almost simple group, with $x \in G$ an order three diagonal automorphism, and $G = \langle S, x \rangle$. Then $C_G(x)$ is even order, unless G and x occur on the following list.

(i) $PGL_3(q)$, $q \equiv 1 \pmod 3$, x in an irreducible torus
(ii) $PGU_3(q)$, $q \equiv -1 \pmod 3$, x in an irreducible torus

Proof.   The almost simple groups containing outer diagonal elements of order three are $E_6(q)_{ad}$, $^2E_6(q)_{ad}$, $PGL_n(q)$ and $PGU_n(q)$, with $n = 3k$. From Lübeck [7], all order three elements in $E_6(q)_{ad}$ and $^2E_6(q)_{ad}$ have even order centralizers.

So assume $G = PGL_{3k}(q)$. Since $x \in G$ is order three, $\hat{x}$ has at most three eigenvalues on the natural module. Hence for $k \geq 2$, $PGL_{3k}(q)$ contains no regular elements of order three. However when $x \in G$ is not regular, $C_G(x)$ contains a copy of $GL_2(q)$, and so in particular is even order. It follows that we may restrict our search to $\langle S, x \rangle = PGL_3(q)$, with x a regular semisimple element of order three.

The outer diagonal elements of order three in $PGL_3(q)$ occur when $q \equiv 1 \pmod 3$, and are contained in either a split or irreducible torus. Since x is regular semisimple of order three, $\hat{x}$ has three distinct eigenvalues $1, \alpha, \alpha^2$ on the natural module, with $\alpha^3 = 1$. In particular, $\hat{x}$ has determinant 1. It follows that if x is contained in a split torus, then $x \in S$ and hence is not an outer-diagonal element.

So we may assume x occurs in an irreducible torus T. Let α be an element of order three in $\mathbb{F}_q^\times$. There is some $\hat{x} \in GL_3(q)$ such that $\hat{x}^3 = diag[\alpha, \alpha, \alpha]$, and $\hat{x}$ has three distinct eigenvalues on the natural module (corresponding to the cube roots of α). In this case, $x \in T$ is regular semisimple of order three, and $C_G(x) = \langle T, w \rangle$ with w an element of order three. Since T has order $q^2 + q + 1$, $C_G(x)$ has odd order. It follows that $PGL_3(q)$, $q \equiv 1$ (mod 3), contains an outer-diagonal element of order three centralizing no involution. Repeating this argument for $PGU_{3k}(q)$ yields that $PGU_3(q)$, $q \equiv -1$ (mod 3), contains outer-diagonal elements of order three with odd order centralizers. $\qquad\square$

5   Proofs of the Main Theorems

Taken together, Sections 2–4 prove Theorem 3. Now recall the statement of Theorem 1.

Theorem 1.   1 Let G be a finite almost simple group, and $x \in G$ be an element of order three. Then x normalizes a nontrivial 2-subgroup of G, unless G and x occur in the following list.

(i) $PSL_2(2^a)$, a odd, x in the unique class of order three elements;
(ii) $PGL_3(2^a)$, a even, x in an irreducible torus;
(iii) $PGU_3(2^a)$, a odd, x in an irreducible torus;
(iv) $^2G_2(q^2)$, $q^2 = 3^{2f+1}$, x in class $(\tilde{A}_1)^3$.

To prove the theorem, it suffices to check whether the list of exceptional order three elements found in Table 1 of Theorem 3 normalize some larger 2-group.

5.1   Alternating and Sporadic Groups

Recall for $n = 5, 6, 7, 9, 10$, Alt_n contains order three elements which centralize no involution. However by an easy inspection all such elements normalize a four group in Alt_n.

Next assume $S = J_3$, and x is in the conjugacy class $3B$. Note that $\text{Aut}(S) \cong J_3 : 2$. By [11], $C_{J_3:2}(x)$ has even order, and there is a single class $2A$ of outer involutions in $J_3 : 2$. Furthermore for $w \in 2A$, $C_{J3:2}(w)$ has order 4896.

Using the list of maximal subgroups of $J_3 : 2$ provided in [3], we find that $PSL_2(17) \times 2$ has order 4896, and that it is the unique maximal subgroup of $J_3 : 2$ with order divisible by 4896. Hence $C_{J_3:2}(w) \cong PSL_2(17) \times 2$. It follows that x is contained in a conjugate of $PSL_2(17)$. But $PSL_2(17)$ has a single conjugacy class of order three elements, and has Alt_4 as a subgroup, so x will normalize a four group in J_3.

5.2   Groups of Lie Type

To begin, let $G = PSL_2(2^a) \cong SL_2(2^a)$. A parabolic subgroup Q of G is a Borel subgroup, and hence has order $2^a(2^a - 1)$. Furthermore, the unipotent radical U of Q is a Sylow-2 subgroup of G. Applying Borel-Tits (see Theorem 26.5 in [8]), the normalizer of any 2-group in $SL_2(2^a)$ is contained in a conjugate of Q. Since $3 \mid 2^a(2^a - 1)$ if and only if a is even, $PSL_2(2^a)$, a odd, contains elements of order three normalizing no nontrivial 2-subgroup.

For $PSL_2(q)$, q odd, we may assume there is a single class of order three elements. When $p \neq 3$, there is a single class, and when $p = 3$ there are two classes, but they are conjugate by an outer automorphism, which preserves the property of normalizing a 2-group. Dickson's Theorem tells us $PSL_2(p) \subset PSL_2(q)$ contains Alt_4 as a subgroup for all odd p, so the normalizer of a four group will contain an element from the class of order three.

For $G = PSL_3(2^a)$ and $PSU_3(2^a)$, the split and partially split tori will be contained in some proper parabolic subgroup Q of G. The unipotent radical U of the relevant parabolic is a non-trivial 2-group, with $N_G(U) = Q$. Hence all elements of order three in these tori will normalize a 2-group.

Next assume $G = PGL_3(q)$, $q \equiv 1 \pmod 3$, with x an element of order three in an irreducible torus. First assume q is even. Then $\hat{x} \in GL_3(q)$ acts irreducibly on three dimensional space, and hence lives inside no parabolic subgroup. Since the normalizer of any 2-group in $GL_3(q)$ is contained in some parabolic, $\hat{x}$ normalizes no nontrivial 2-group in $GL_3(q)$. It follows that x normalizes no nontrivial 2-subgroup in $PGL_3(2^a)$, a even. The same argument shows an element of order three in an irreducible torus of $PGU_3(2^a)$, a odd, will normalize no nontrivial 2-group.

Now assume q is odd. The order of $GL_3(q)$ is $q^3(q - 1)^3(q^2 + q + 1)(q + 1)$, and the order of a split torus T in $GL_3(q)$ is $(q - 1)^3$. Note $3 \mid q^2 + q + 1$, but $9 \nmid q^2 + q + 1$. It follows that $N_{GL_3(q)}(T) = T \rtimes Sym_3$ contains a Sylow-3 subgroup of $GL_3(q)$. In particular, if K is a maximal elementary 2-subgroup of T, then $N_{GL_3(q)}(K)$ contains both T and an additional order three element permuting elements in K. So $N_{GL_3(q)}(K)$ contains a Sylow-3 subgroup of $GL_3(q)$. By viewing the images of K and T in G, it follows that all order three elements in $PGL_3(q)$ will normalize a 2-subgroup. The same type of argument applies to elements of order three in $PGU_3(q)$, q odd, $q \equiv -1 \pmod 3$.

Next let $G = PSL_3(3^a)$, and $\hat{x} = J_3$. Again, a split torus of $SL_3(3^a)$ contains an elementary abelian 2-group K, with

$$y = \begin{bmatrix} 0 & 0 & 1 \\ 1 & 0 & 0 \\ 0 & 1 & 0 \end{bmatrix} \in N_{SL_3(3^a)}(K).$$

Since y has Jordan form J_3, x will normalize a 2-group in G. Restricting to the subgroup $SL_3(3^a) \subset PSL_4(3^a)$ occurring on the J_3 block, the same argument can be used to show that x normalizes a 2-group of $PSL_4(3^a)$, when $\hat{x} = J_3 \oplus J_1$. This argument can also be repeated to show that x normalizes a nontrivial 2-group in $PSU_3(3^a)$ and $PSU_4(3^a)$ in the relevant cases.

Now consider $G = G_2(q)$, $q = 3^f$, with x in the class $(\tilde{A}_1)^3$. There is a subgroup of type $A_1 A_1$ in G (which is the centralizer of an involution). One of the factors is generated by two short root subgroups and the other by two long root subgroups (see [8], §13). The order three element $x \in (\tilde{A}_1)^3$ is a diagonal product of a long and short root element. Each A_1 contains a copy of Alt$_4$, so each long and short root element will individually normalize a four group K, and hence x will normalize $K \times K$ in Alt$_4 \times$ Alt$_4$.

Finally consider $G = {}^2G_2(q^2)$. G has order $q^6(q^6 + 1)(q^2 - 1)$, $q^2 = 3^{2f+1}$, so its Sylow-2 subgroups have order 8. One of its Sylow-2 subgroups P will be contained in ${}^2G_2(3) \cong P\Gamma L_2(8)$. Note that up to conjugacy, $P\Gamma L_2(8)$ contains a single class of involutions, and a single four group K. It follows that ${}^2G_2(q^2)$ will share this property.

Let y be an element from the class of involutions in $P\Gamma L_2(8)$. By **Theorem 3**, $N_{P\Gamma L_2(8)}(\langle y \rangle)$ has order prime to three. Furthermore, $N_{P\Gamma L_2(8)}(K)$ has order $2^3 \cdot 3$, and $N_{P\Gamma L_2(8)}(P)$ has order $2^3 \cdot 3 \cdot 7$. So, up to conjugacy, there is a single class of order three elements in the normalizer of any 2-group. However, there are two classes of order three elements in G. So ${}^2G_2(q^2)$, $x \in (\tilde{A}_1)^3$, yields a final class of exceptions.

5.3   General Finite Groups

We have now classified all order three elements in the almost simple groups which normalize no nontrivial 2-subgroup. This enables us to prove the following general theorem concerning finite groups. Note when $O_3(G)$ is nontrivial, it is quite messy to pin down necessary conditions for the existence of order three elements normalizing no nontrivial 2-subgroup.

Theorem 2.   Assume that G is a finite group with components $L_1, ..., L_n$, and $O_3(G) = 1$. Let $x \in G$ be an order three element normalizing no nontrivial 2-subgroup of G. Then *(i)* if N is any normal subgroup of G with order prime to three, then N has order prime to six;

(ii) x normalizes every component L_i of G;

(iii) The image of $\langle L_i, x \rangle$ in $Aut(L_i)$ appears on the list of exceptions in **Theorem 1**, for every component L_i of G.

Proof.   In the following assume $O_3(G) = 1$, and $x \in G$ is an order three element normalizing no nontrivial 2-subgroup.

(i) Assume N is a normal subgroup of G with order prime to three. Furthermore, assume six divides the order of N, and P is a Sylow-2 subgroup of N. Then by the Frattini argument $G = N N_G(P)$. Since N has order prime to 3, a conjugate of x will normalize P. This contradicts our assumption that the class of x normalizes no nontrivial 2-group. So we may assume N has order prime to six.

(ii) Assume $x \in G$ does not normalize some component L_i of G, so that L_i, L_i^x and $L_i^{x^2}$ are distinct commuting components. Let w be an involution in L_i. Then $y = w \cdot w^x \cdot w^{x^2}$ is an involution in G, and $yx = xy$.

(iii) Pick some component L_i, where L_i is a perfect central extension of the simple group S_i. Since x normalizes L_i, $\langle L_i, x \rangle \subset N_G(L_i)$. Furthermore, $Aut(L_i)$ can be embedded in $Aut(S_i)$. So the image of $\langle L_i, x \rangle$ in $N_G(L_i)/C_G(L_i) \subset Aut(L_i) \hookrightarrow Aut(S_i)$ must appear on the list of exceptions to **Theorem 1**. $\square$

Note while **Theorem 2** gives necessary conditions for the existence order three elements which normalize no nontrivial 2-group, it does not claim these conditions to be sufficient. Unfortunately, no version of $(i) - (iii)$ will yield such conditions. For instance, if $G = (\mathbb{Z}_5 \times \mathbb{Z}_5) : \text{Alt}_4$, then G satisfies $(i) - (iii)$, but the single class of order three elements in G normalizes a four group. In some sense this counterexample is generated by the occurrence of an order prime to six normal subgroup in G. However, even under the stronger assumption that no such prime to six normal subgroups appear, the conditions would still not be sufficient.

For instance, let x be in the single class of order three elements in $G = PSL_2(2^4)$. Let $H = G \wr \text{Sym}_2$, where Sym_2 acts by permuting the copies of $PSL_2(2^4)$. Then $y = (x,x) \in H$ is order three, y, H satisfy conditions *(i)–(iii)* above, and H contains no order prime to six normal subgroups. However y centralizes the involution swapping the two copies of $PSL_2(2^4)$. Hence there is no obvious way to extend our characterization of order three elements in the almost simple groups which normalize no non-trivial 2-subgroup to a complete description of such elements in more general classes of finite groups, even under the assumption that $O_3(G) = 1$.

Acknowledgements

The author would like to thank Geoff Robinson for suggesting the question considered in this paper, and Robert Guralnick and Gunter Malle for helpful comments and suggestions on an earlier version of this manuscript. This research was partially funded by NSF grant DMS-1265297.

References

[1] A. Borel, *Linear Algebraic Groups*, W.A. Benjamin, Inc., New York, 1969.

[2] R. Brauer, *Representations of Finite Groups,* Lectures on Modern Mathematics, vol. I, Wiley, New York, 1963, 133–175.

[3] J. Conway, R, Curtis, S. Norton, R. Parker and R Wilson, *ATLAS of Finite Groups: Maximal Subgroups and Ordinary Characters for Simple Groups*, Oxford University Press, Oxford, 1986.

[4] D. Gorenstein, R. Lyons, and R. Solomon, *The Classification of the Finite Simple Groups*, Number 3, Mathematical Surveys and Monographs, Vol. 40, No. 3, American Mathematical Society, Providence, R.I. 1998.

[5] J. Humphreys, *Conjugacy Classes in Semisimple Algebraic Groups*, Mathematical Surveys and Monographs, Vol. 43 American Mathematical Society, Providence, R.I. 1995.

[6] M. Liebeck and G. Seitz, *Unipotent and Nilpotent Classes in Simple Algebraic Groups and Lie Algebras*, Mathematical Surveys and Monographs, Vol. 180 American Mathematical Society, Providence, R.I. 2012.

[7] F. Lübeck, Online data for groups of Lie type, (http://www.math.rwth-aachen.de:8001/Frank.Luebeck/chev).

[8] G. Malle and D. Testerman, *Linear Algebraic Groups and Finite Groups of Lie Type*, Cambridge University Press, Cambridge, 2011.

[9] G. Robinson, The number of p-blocks with a given defect group, *Journal of Algebra*, 84:2 (1983), 493–502.

[10] G. Robinson, Involutions, weights and p-local structure, *Algebra and Number Theory*, 5:8 (2011), 1063–1068.

[11] R. Wilson, P. Walsh, J. Tripp, I. Suleiman, R. Parker, S. Norton, S. Nickerson, S. Linton, J. Bray, R. Abbott, *ATLAS of Finite Group Representations - Version 3*, (http://brauer.maths.qmul.ac.uk/Atlas/v3/).

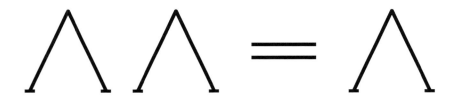

Catherine Christer Hennix
B/W Equation from *Algebra w/ Domains*
1973–91

Domains of Variation
133

1

The art and writings of C.C. Hennix draw extensively upon certain kinds of mathematical formalism, most prominently category theory and categorical logic, but also set theory, type theory, L. E. J. Brouwer's intuitionism, intuitionistic logic, Alexander Esenin-Volpin's ultrafinitism, and David Hilbert's proof theory (or her interpretation of it). In particular, her published writings of the 1970s and '80s, works like *The Yellow Book* (1989) and *Notes on Toposes and Adjoints* (1976), contain a vast array of mathematical machinery presented within an even larger kaleidoscope of references, including early analytic philosophy, Eastern thinking, Japanese Noh drama, formal linguistics, and other topics.

While these writings were not intended as systematic, and have an avowedly esoteric and fragmentary quality, it is interesting to consider what attracted Hennix to the mathematical concepts she sought to employ, and how these references relate to her better-known output as a composer and artist. It is a curious and almost characteristic feature of these works that in spite (or perhaps because) of their richly variegated standpoint, they make little effort in the way of explaining themselves. Their poetic, formal, and collage-like character lends them an introspective or inward-looking quality, which both mirrors and obscures the meanings of the texts. This paper aims to give some clarity to those writings, providing an ambient theory that encompasses Hennix's technical, aphoristic, and often speculative prose.

Although these texts resist any complete interpretation, it is instructive to place them in the context of her substantial artistic engagements, with the music of La Monte Young and Pandit Pran Nath, and her ongoing collaborations with Henry Flynt. One could choose to adopt a technical viewpoint and evaluate specific claims in the texts. In the end, however, it is more fruitful to paint a somewhat broader portrait, one of a highly original artistic and intellectual milieu arising out of an unlikely confluence of mathematics and the 1960s avant-garde. While there are several ways of approaching this task, I will focus first on Hennix's interest in category theory as it relates to her engagement with the music of Young, and second on the act of semantical interpretation as it pertains to her ongoing dialogue with Flynt.

2

As is clear from the first paragraph, Hennix's primary mathematical interests lie in logic, and more specifically nonstandard foundational positions for mathematics. While the viewpoints she drew upon seldom became widely adopted (and in some cases are almost completely unknown within the mathematics community), in a more general sense they do intersect with important (and non–philosophically motivated) developments in twentieth-century mathematics. Any substantive foundational theory should arise out of mathematical practice, and indeed set theory, category theory, and type theory developed out of significant problems in analysis, topology, and the theory of computation. Furthermore, as we shall discuss later, the notion of a topos, a central concept in her work and one to which she ascribes a great deal of philosophical importance, initially arose as a tool for tackling important problems in topology and algebraic geometry.

When convenient, I will adopt a more general mathematical viewpoint for ease of exposition, even if this standpoint isn't explicitly advocated in the writings. Indeed, while her use of a topos often hews closely to the logical conception of a "universe of sets," the topological perspective may in fact be more relevant to her underlying interests.

To begin, I will provide some general mathematical background necessary to make headway into the writings. One could go a ways in this direction, but I will restrict discussion to the most pertinent concepts: a description of the classical continuum, Brouwer's notion of a choice sequence, and some background in the development of category theory.

3

One initial point of confusion (and sometimes terminal boundary) surrounding Hennix's work is that she uses mathematical terminology for nonmathematical purposes or assigns it extra-mathematical meanings. For instance, besides being a category satisfying certain axioms, in these writings a topos often refers to a "modality of thought" or "state of consciousness."

In fact, much of the mathematical formalism Hennix adopts is at some point or other reinterpreted in a not-fully-defined language of introspection. For instance, she often views Brouwer's Creative Subject as an uninterrupted stream of sensations to which "mental operations" may be applied. In *Excerpts from Parmenides and Intensional Logics*, she suggests that "the minds mental operations are: ATTENTIONS (arrows, sets), COLLATIONS ($=$, $\neq$), CONNECTIONS (CON, θ_ω), MEMORY/RECOLLECTIONS ($\mathcal{I}_\omega$), PERCEPTIONS ($\wedge_\omega$), and INTENSIONS ($\vee_\omega$)" before resuming a poetic dialogue surrounding "Mind, upon entering the domains of the Metaxyan Galaxy, in communion with Kosmos."[1] In many cases, philosophically suggestive or poetic couplings are made between widely different arenas, but these leaps are indicated within the mathematical formalism at hand. As any connection between these pairings cannot be expressed in the language in question, some confusion arises as to the intent of using formal languages, adding to the esoteric quality of the texts. While the precision of mathematical language has a poetic appeal, yielding connotations that might not appear in more expository contexts, it seems that the arguments, given their emphatic nature, would be better expressed in a natural language setting. Indeed, Myles Tierney (one of the founders of topos theory) supposedly remarked that Hennix had succeeded in reintroducing all the vagaries in the subject that he and F.W. Lawvere had spent their professional lives trying to remove.

On the other hand, it is not difficult to locate Hennix's approach in certain mathematical precedents. For instance, the notion that mathematics can be used to formalize a mode of thinking has several highly interesting precursors. One aspect of Brouwer's intuitionistic program was a mathematical description of a process of introspection (although it should be emphasized that Brouwer developed a complete and rigorous mathematical theory). Similarly, at various points Hilbert claimed his proof theory was "nothing else but a formalization of our process of thought." Going further in this direction, in *What Are Numbers and What Should They Be?* Richard Dedekind "proves" the existence of infinite systems by devising a bijection between a collection of thoughts and one of their proper subsets. In this case, the *existence* of an infinite system is justified by an appeal to a process of introspection. It is in this context that one can locate Hennix's attempts to wed mathematical formalism and internal or introspective states.

Furthermore, one can trace her specific interest in topos theory to her time spent at the University of California, Berkeley, as an exchange student. In 1971, twenty-three-year-old Hennix arrived at one of the leading centers of logic research in the world, Berkeley's Group in Logic and the Methodology of Science. By this time, she had already composed several electroacoustic compositions at Elektronmusikstudion Stockholm (EMS), one of the world's leading centers for electronic music, which had been anthologized and performed at major museums. She had met and begun work with La Monte Young (and in 1970 realized one of his *Drift Studies* using the new phase-locked oscillators at EMS), and was an active jazz drummer who would soon appear with Don Cherry on the soundtrack to Alejandro Jodorowsky's *The Holy Mountain* (1973). At the same time, she was a student of linguistics, philosophy, and mathematics at Stockholm University. While at Berkeley, besides working on logic, she began her extensive raga studies with Pandit Pran Nath, who was teaching at Mills College in Oakland at the invitation of Terry Riley. Let us now set the scene of Bay Area mathematical logic at the time of Hennix's arrival.

4

In 1963, the Stanford mathematician Paul Cohen introduced his method of forcing, a revolutionary technique for constructing new models of set theory. He amazed the mathematical world by using this method to prove the independence of the continuum hypothesis from the standard axioms of set theory (called ZFC), later winning a Fields Medal for this effort. However, a new method can be more valuable than even the most substantial result, and there was a concerted effort within the logic community to understand this new construction.

By the mid- to late 1960s, Robert Solovay and Dana Scott (who were then teaching at Berkeley and Stanford) reformulated Cohen's approach in algebraic terms, in the form of "Boolean-valued models" of set theory. Shortly thereafter, two prominent category theorists, Lawvere and Tierney, interpreted these Boolean-valued models in terms of categories of sheaves on Boolean algebras, in effect expressing them in the language of category theory.

In 1966, Lawvere encountered Alexander Grothendieck's notion of a topos, which generalized the notion of a sheaf on a topological space. Lawvere and Tierney saw foundational significance in this notion and axiomatized the related concept of an elementary topos in 1969–71. Almost immediately, the "Boolean-valued models" of Solovay and Scott (or their categorical expressions) were seen to satisfy the axioms of a topos, which in turn allowed much of Cohen's original forcing technique (and in particular the construction of new models of set theory) to be translated into the language of category theory. From a logical perspective, a topos could now be viewed as a "universe of sets," although initially it was conceived as a vastly generalized topological construction.

This was state-of-the-art work in logic when Hennix arrived at Cal. One can detect in this setting a longstanding and pervasive influence on her output. Indeed, Solovay, who was then a young faculty member at Berkeley and had recently completed his PhD under Saunders Mac Lane (one of the founders of category theory) at the University of Chicago, was a key contributor to all of these developments, and his work, along with that of the constructive proof theorist Georg Kreisel at Stanford, seemed to exert a particularly strong influence on Hennix.

Locally, many of the notes on category theory appearing in *The Yellow Book* and *Notes on Toposes and Adjoints* were compiled from discussions with students and faculty during her time at Berkeley. Her compositions of this period, pieces including *Forcing (1973)*, *Fixed Points* (1973), and $\Box_\kappa$ (1976), are expressed entirely in the language of set theory and deal with Boolean-valued models, large cardinals (infinite sets with size much greater than what is typically required to do mathematics), and other topics she encountered as a student. Furthermore, the construction of the famous Solovay model in 1970, which uses forcing and the existence of a large cardinal to construct a model where the classical continuum has nicer properties than is typical (every subset is Lebesgue-measurable), seems an interesting precursor to Hennix's later repeated attempts to find new models for a "continuum of [musical] intervals," as she proposes in *Forcing*.

One can detect how Hennix, through her formative experiences as a student, arrives at certain central themes in her work. However, this does not necessarily explain the intent of her project or illuminate the connections she was trying to draw. Given the fragmentary, formal, and most commonly poetic nature of the writings, it is not reasonable to expect a systematic answer to this question. Nonetheless, it is instructive to look towards the mathematical concepts Hennix employs. In general terms, they relate to formalizations of continuous phenomena, or intensional theories (based on the senses of terms) rather than extensional ones (which concern the abstract existence of objects).

In many cases, the mathematical formalism she adopts (intuitionism, topos theory, type theory, etc.) is concerned with expressing continuity as a primitive notion, rather than as an additional layer of structure added upon a discrete or extensionally defined collection of objects. This experiential notion of continuity provides a central undercurrent to her writings, and she likely saw mathematical language as an exact tool for delineating this concept. Indeed, it is precisely this synthetic process that Brouwer sought to characterize with his intuitionistic continuum.

Although never explicitly stated in these terms, this is also almost certainly the connection Hennix wished to establish between mathematics and her other artistic activities. In particular, it is exactly this kind of synthetic construction that underlies Young's conception of "tuning as a function of time," and several facets of his compositions from the 1960s onwards. Similarly, one feature of Flynt's early concept art was the development of new formal systems for introspective experiences. For instance, in *Concept Art Version of Mathematics System 3/26/61* (now titled *Illusions-Ratios*) (1961/3), "sentences" of the formal language are taken to be perceived length-to-width ratios of the logical symbol "⊥" (hence the "syntax" of the language is composed of synthetic experiences), and deductions are understood as reorderings of these continuous phenomena. These pieces provide a "logic" of the mind's presentational powers, which proved influential on Hennix.

However, Flynt's aims were somewhat different than Hennix's, and he was inspired by different philosophical source material, in particular Rudolf Carnap's *The Logical Syntax of Language* (1934) and Ernest Newman and James Nagel's *Gödel's Proof* (1958). Flynt began as an undergraduate mathematics student at Harvard in the fall of 1957 (in the class ahead of Tony Conrad, Saul Kripke, and Ted Kaczynski). During this time, Flynt sampled course offerings widely, including W. V. Quine's mathematical logic course (both he and Conrad struggled with the final exam, although a drawing Flynt made during class is now in the collection of MoMA) and real analysis and algebraic geometry courses with John Tate, before eventually withdrawing from school in the spring of 1960.

Through his coursework, Flynt became interested in formalism as a foundational position for mathematics. Roughly put, formalism reduces mathematics to syntactic manipulation, or the mathematical manipulation of strings of symbols. However, this syntax may be interpreted on models, which in turn tells you information about structures you are interested in. Flynt perceived a connection between this combinatorial syntactic approach and things like process art and indeterminacy, in particular in the work of Richard Maxfield and John Cage.

In the summer of 1960, Flynt ventured into (in his own phrasing) "unexplored regions of formalist mathematics" by developing purely syntactic formal systems which, in his mind, precluded any notion of a model or interpretation—and hence "meaning" in a traditional mathematical sense.[2] He saw this project, which he titled concept art, as related to mathematical formalism but without the imputed metamathematical significance. Indeed, by binding syntax to perceptual experience (in particular to nonstable perceptual states), Flynt believed he had found a critique of formalism, which he associated with a criticism of mathematics more broadly.

Flynt's "meaningless" logical systems in turn led him to more general conceptions of "veramusement" and "brend," which, in logical terms, concern a free play of the mind's presentational powers. While Flynt had a specific logical paradigm in mind, he also considered his approach from the perspective of new music, in particular the work of Cage. Cage's attitude toward composition concentrated the "work" in the mental states of the viewer. This proved a significant influence on Flynt's concept art, and later Hennix's epistemic art. Indeed, Flynt claimed concept art was an attempt to "apply new music to metamathematics."[3] Although Cage's focus on a free play of the mind's presentational powers was conceived in compositional terms, one might also locate it in classical aesthetics; his notion of "purposiveness without a purpose" is derived (likely via John Dewey) from Kant's theory of aesthetic judgment. We will discuss this connection at greater length later on.

Through Young, Hennix became aware of Flynt's work, and they have been close collaborators since the mid-1970s. In many instances, one can sense Hennix adopting a semantical viewpoint on Flynt's efforts. In his construction of logics of subjective experiences, Flynt follows Carnap's dictum: "In logic, there are no morals. Everyone is at liberty to build up his own logic, i.e., his own form of language, as he wishes."[4] Flynt introduces syntactically complex languages (indeed, it is often impossible to precisely specify what the syntax is, given its correlation with a continuum of experiences), where the "logical terms" (being a free play of cognitive powers) preclude any external interpretation.

However, instead of an uninterpreted "logic" of introspective experiences, from a semantical standpoint one could also propose introspective "interpretations" or "nonstandard models" of mathematical formalizations, as Hennix frequently does in *The Yellow Book*—for instance, in her association of toposes with different "modalities of thought." In this way, while Flynt's concept art could be considered "syntax rich, semantics poor," in "Brouwer's Lattice," Hennix proposes her work as being "syntax poor, semantics rich," adding, "Let the Diagrams Commute," appropriating then-popular slogans from category theory.[5] Let us now consider some additional mathematical background necessary to evaluate these efforts.

6

In a classical framework, there is an underlying assumption that mathematics concerns constant structures, even in its attempts to model motion in space or movement in time. Hence set theory, with its primitive notions of elements and membership, became the presumptive foundational position for mathematics, and Richard Dedekind's "arithmetization of geometry" provides the standard mathematical account of the continuum. From this perspective, mathematics begins with bare sets to which structure, often of a "spatial" variety, is added. One consequence of this analytic viewpoint is that, at a foundational level, the world is viewed principally in discrete or disconnected terms.

The classical continuum provides a striking example of this vantage point. A line is perhaps the most immediately given and intuitable spatial entity, and serves as a common model for both time and space. In standard mathematical practice, the continuum is composed of discrete elements that, taken together, form a single continuous whole. Of course, the way in which these atomistic pieces might fit together to form a continuous whole is one of the most vexing problems in the history of philosophy and mathematics. As Leibniz notes, "There are two labyrinths of the human mind, one concerning the composition of the continuum, and the other concerning the nature of free will, and they both arise from the same source, infinity."[6]

For future comparison, let us briefly suggest a standard mathematical construction of the continuum. The set of natural numbers $\mathbb{N} = \{0, 1, 2, ...\}$ and the set of integers $\mathbb{Z} = \{..., -2, -1, 0, 1, 2, ...\}$ can be successively enumerated, and come equipped with a natural ordering " $<$." From the integers, one can construct the set of all fractions by considering ratios of integers up to fractional equivalence (i.e., $1/2 = 2/4 = 4/8 = ...$). Together with the natural ordering on the set of all fractions, this yields the rational numbers $(\mathbb{Q}, <)$. The rational numbers are dense (in between any two fractions there exists another fraction), but "gappy" in the sense that one can define sequences of fractions that converge to no rational number (such as a sequence of fractions that get closer and closer but never reach $\sqrt{2}$). These gaps appear naturally and frequently. For instance, some gaps arise as solutions to simple algebraic equations like $x^2 = 2$. All such gaps are called *irrational numbers*.

The classical continuum is constructed as a "completion" of the rational numbers, by filling in all missing gaps (such as $\sqrt{2}$, π, e, etc.). Together with an ordering, this yields the real numbers ($\mathbb{R}$, $<$), which serve as the standard model of the continuum. Note in this procedure, a line is associated with "real numbers," which can be expressed in terms of sequences fractions (and thereby reduced to integers and ultimately natural numbers). Hence this process is often referred to as an "arithmetization of geometry."

One curious feature of this completion is that there are many, many more missing irrational numbers than there are fractions. In fact, while there is a natural way to enumerate all fractions, there is no listing of the missing irrational numbers, even in theory. While intuitively it seems that a few gaps are being filled, in reality the probability of picking a random point on the classical continuum and it being a fraction is zero.

Hence, although each classical real number exists as an independent entity, each with its own place and determined with infinite precision on a line, nearly all of these objects are ineffable in the sense that, taken together, they vastly outnumber the language used to describe them. As a result, this "completion" must necessarily occur in some indirect fashion. At best, almost all of the uncountable infinity of missing irrational numbers can be described as infinite nonrepeating decimal expansions, and approximated through finite initial segments of this expansion.

However, knowing an expansion up to any finite stage does not tell us what will happen in the future, and reducing a real number to a breadthless point requires infinite precision with respect to its placement on the line. In this respect, the missing irrational numbers may be viewed as "actually infinite objects"; they can be identified with equivalence classes of infinite sequences of fractions approaching but never reaching the missing gap. But it necessarily takes a full infinite sequence (or function describing this sequence) to completely specify an irrational number. The completion of the continuum can be seen as an abstract process of identifying "points" with these "infinite objects." Furthermore, this construction can never be fully explicit in the sense that there are too many missing points to describe.

7

In general terms, topology studies space up to continuous deformation. While one is typically interested in properties of a particular space, in practice it is often beneficial to consider continuous mappings in and out of a space in determining properties of the space itself. This was one of the basic perspectival shifts in mathematics of the twentieth century, as Jean-Claude Pont (via Colin McLarty) notes on the Cantor-Dedekind correspondence: "One can trace the origin of modern topology to the discovery that mappings which transform one manifold into another teach us as much about the manifolds as do the manifolds themselves."[7] Hence, while abstractly topology studies the coherence of space, at a more practicable level it examines properties preserved by continuous functions between spaces. Indeed, topological objects themselves are frequently identified with such mappings. For instance, a knot is often viewed as a continuous function from a unit circle into three-dimensional space.

It is worth noting that unlike the set-theoretic approach of characterizing objects as sets satisfying certain properties, from a topological viewpoint the ambient space of an object is in some sense "built in" to the object itself. A function from a circle of radius one to 3-space is different than a function from the unit circle to a sphere, just as a knot in three-dimensional space is somehow different than a knot appearing on a sphere. This emphasis on maps between spaces over the spaces themselves is often taken as a starting point for the development of category theory.[8]

However, even after adopting the "functional" viewpoint of modern topology, set-theoretic assumptions still underlie the standard formulation of continuity. A topological space is specified by taking a set X and assigning certain subsets of X as *open* sets (which intuitively correspond to continua, or unions of such regions). Typically, some collection of basic open sets are given, from which all others can be formed. For topological spaces X and Y, continuous functions between X and Y are then defined in terms of preserving properties of open sets (the inverse image of an open set is open). Hence a discrete collection is given first, to which a spatial structure (in this case a topology) is added on top as an additional layer. As noted earlier, this "spatialization" of discrete objects is ubiquitous in mathematical practice: a group becomes an algebraic or topological group by adding a topological structure compatible with the group operation, a collection of abelian groups is parameterized to form a sheaf, and so on.

For the classical continuum, one common basis of open sets is all intervals of the form (q, r) where q and r are fractions (in other words, the set of all real numbers between q and r without including the endpoints q and r themselves). The complement of an open set, i.e., the set of all points not contained in the open set, is called a *closed* set. For instance, in the classical continuum, any singleton set $\{s\}$ is closed, intervals of the form $[q, r]$ that contain the endpoints points q and r are closed, etc. To turn an open set (and indeed any set) into a closed set, one adds all missing "boundary" or "limit" points by taking a closure, for instance by adding the endpoints q and r to (q, r). From this perspective, the classical continuum is the completion or topological closure of the rational numbers, as every irrational number is the limit point of a sequence of rational numbers.

Finally, taking appropriate unions, intersections, and complements of all basic open sets gives a σ-algebra. In the case of the classical continuum, this yields the Borel sets, which can be viewed as the "describable" subsets of the continuum. Although a line appears placid and unassuming, general subsets of the classical continuum are known to have quite pathological properties, and indeed are hard to discuss at all. The Borel sets are relatively well behaved and allow one to define intuitive notions like length and size. For general topological spaces, the algebra of open sets will play an important role later on.

8

L. E. J. Brouwer was one of the most distinguished topologists and mathematicians of the first half of the twentieth century. Perhaps given his topological inclinations, he was dissatisfied with a reduction of the world to discrete terms. Starting as early as his dissertation of 1907, Brouwer suggested the continuum was mischaracterized by a reduction to completed atomistic points. Although he is now best known for his fundamental fixed-point and dimensionality theorems (two of the most important results of early-twentieth-century mathematics), the composition of the continuum was perhaps the most recurrent theme in his work, and a topic he continually revisited throughout his life. He writes: "The continuum as a whole is intuitively given to us; a construction of the continuum, an act which would create all its parts as individualized by mathematical intuition is unthinkable and impossible. The mathematical intuition is not capable of creating other than countable quantities in an individualized way . . . the natural numbers and the continuum are two aspects of a single intuition (the primeval intuition)."[9] Here Brouwer objects to the size of the continuum required to speak of it as individualized points, noting that any "fixed" and discrete reduction betrays both our mathematical intuition and the faculties of our imagination.

While many prominent mathematicians (such as Émile Borel, Sophus Lie, Hilbert, Hermann Weyl, and others) had expressed doubts about the arithmetization of geometry, Brouwer was the first to propose a new alternative mathematical theory of the continuum, one based on a primitive notion of coherence and "stickiness." The mathematical framework needed for this new theory is called intuitionism.

While it is often thought that intuitionism restricts the rules available to do math, and hence provides a weaker theory of mathematics (Hilbert compared it to proscribing a boxer the use of his fists),[10] in reality it is incomparable to classical math and in many ways strengthens it, in the sense that nice properties that are false in classical analysis are true in Brouwer's theory. For instance, it is provable that every full function on the intuitionistic continuum is continuous. Conversely, many theorems of classical mathematics, and indeed his own fixed-point theorem, are false in intuitionistic analysis.

Of course, one should expect and indeed hope that two fundamentally different conceptions of the continuum would lead to different mathematical theories. Furthermore, once the differences in conception are accounted for, corresponding "approximate" versions of many classical theorems that fail on the intuitive continuum can be shown to hold given the appropriate statement (the intermediate value theorem, his fixed-point theorem, etc.).

It should be stressed that Brouwer's alternative continuum is a rigorous mathematical theory. While philosophers going back to Aristotle (and earlier) speculated on the coherence of the continuum, Brouwer proposed a precise mathematical framework modeling one possible viewpoint on these phenomena. Out of this effort came a highly original conception of the relationship between the discrete and the continuous, a model for the act of doing mathematics, and a mathematical theory of introspection. In total, it was one of the most original conceptual developments of the twentieth century, and one that overlaps with other fundamental artistic achievements in a manner not yet fully appreciated.[11]

To understand the uniqueness of Brouwer's contribution, the passage cited above bears further reflection. Following Kant, Brouwer believed the basic irreducible unit of math was a movement of time. While sensations may come and go without producing any noticeable reaction, mathematics proper begins with an act of attention. As Dirk van Dalen nicely summarizes, sensations pass but "at a certain point the subject may actively interfere as follows: the subject notes, or fixes his attention on a sensation, which is subsequently followed by another sensation. The first sensation is stored in memory at its passing away, being replaced by another sensation. The result is that the subject is aware of both the remembered and the present sensation."[12] Brouwer refers to this act as a "two-ity," or "the falling apart of one life-moment to the next."

Of course, for any given "two-ity" there are specific qualities associated with each sensation. Through introspection, one can abstract away from these properties. Brouwer writes, "Mathematics is a languageless activity of the mind having its origin in the basic phenomenon of a move of time . . . which is the falling apart of a life moment into two distinct things. . . . If the two-ity thus born is divested of all quality, there remains the common substratum of all two-ities, the mental creation of the empty two-ity. This empty two-ity and the two unities of which it is composed, constitute the basic mathematical systems."[13] From this fundamental act of introspection, the awareness of time passing, one may construct the natural numbers (first as a two-ity, then a three-ity, a four-ity, and so on), and from this the integers, the rationals, infinitely proceeding mathematical systems, and infinitely proceeding sequences of mathematical systems previously acquired.

There is a subtle but crucial distinction between these systems and their classical counterparts. Rather than existing as completed sets, these constructions occur in time, yielding potentially infinite or infinitely proceeding structures. Hence in the set of natural numbers $\mathbb{N} = \{0, 1, 2, ...\}$ the symbols "..." have a somewhat different meaning than they do in their classical counterpart, in the sense that $\mathbb{N}$ does not refer to a completed mathematical object but rather a process of generation. All relevant constructions are both time- and subject-dependent, and potentially infinite structures are dynamic.

Although it is not immediately obvious, Brouwer's conception of a two-ity also leads to an interesting resolution to the discrepancy between the discrete and the continuous. He writes, "Ur-intuition of two-ness (two-ity): The intuitions of the continuous and the discrete join here, as the second is thought not by itself, but under preservation of the recollection of the first. The first and the second are thus kept together and the intuition of the continuous consists in this keeping together (*continere* = keeping together). This mathematical ur-intuition is nothing but the contentless abstraction of the sensation of time. That is to say, the sensation of 'fixed' and 'floating' together or of remaining or changing together."[14] In other words, between every distinct two-ity, there is an entire continuum of events, namely the continuous movement of one moment of time to the next. Hence numbers and the continuum are given to us simultaneously and inseparably, one as the observation of time passing and the other as what happens between these two moments. This is the meaning behind Brouwer's comment that "the natural numbers and the continuum are two aspects of a single intuition."

What happens between two acts of attention cannot be reduced to a collection of atomized points. It is not "exhaustible by the interposition of new units and therefore can never be thought of as a mere collection of units."[15] Instead, this continuum is constituted through *choice sequences*: never-finished creations developing in time according to an idealized mathematician's attention. Some choice sequences (the lawlike ones) are completely determined and can be thought of as explicit algorithms, while others (lawless sequences) are generated quite freely by the subject. (It is a subtle distinction, but while in descriptive set theory real numbers are often associated with completed infinite branches of a tree called a Baire space, choice sequences may be viewed as the finite branches of this tree as they develop in time.)

Irrational numbers were previously seen as "actually infinite" or (with some abuse of notation) "transcendental" objects, in the sense that they were identified with infinite sequences of fractions approaching them, or placed with infinite precision on a line. Choice sequences, on the other hand, are always potentially infinite, and contain a slight halo of "blurriness" in the form of a tail that is never fully resolved. It is never precisely clear where they are, and there is always a small (but clearly defined) amount of wiggle room for where they will head in the future. Hence in this setting a real number is divested of its abstract discreteness. Instead, in Brouwer's conception, one sequentially chooses nested closed intervals of the form $\lambda_n = [\ ^a/2^n, \ ^{a+2}/2^n]$, $a \in \mathbb{Z}$,

$n \in \mathbb{N}$. In particular, the continuum as a whole is not a fixed creation to which dynamic properties are added as an additional layer of structure; it is in its fundamental nature dynamic.

Mathematically, the discrepancy between the intuitive and classical continuum is expressed in terms of Brouwer's weak-continuity principle, which asserts that well-defined properties of choice sequences are determined at some finite stage of development.[16] From this principle, it can be shown that no natural ordering "<" exists on the intuitive continuum, as it does for the classical continuum $(\mathbb{R}, <)$. For instance, it is provable that there exist choice sequences x such that "$x < 0$ or $x = 0$ or $x > 0$" is false. This, of course, matches the informal description of blurriness given in the preceding paragraph.

The continuity principle concisely expresses the strong coherence and connectedness of the intuitive continuum. As van Dalen notes, "The continuum of Brouwer shows a strong 'syrup-like' behaviour; one cannot cut up the continuum into two parts, because in the act of cutting the status of a large number of reals is left open, e.g. suppose one wants to cut the continuum at the origin, then there are lots of points for which $a > 0$ or $a < 0$ or $a = 0$ is unknown."[17] Brouwer explicitly produced such "underdetermined real numbers" as so-called "weak counterexamples." Unlike its classical counterpart, the Brouwerian continuum cannot be reduced to any underlying static framework or constant structure. It expresses the primordial coherence of the intuitively given continuum, formulated in a rigorous mathematical setting.

9

By the 1940s, many deep theorems in topology (such as Brouwer's fixed-point theorem) had been reformulated in algebraic terms, specifically in the language of homology and its close relative, cohomology. Several homological theories existed by then, each with their own distinctive character and results. These methods associated algebraic structures such as groups to topological spaces, in effect producing "algebraic images" of topological spaces.

In general, topological spaces are unstructured and difficult to distinguish. For instance, it is not easy to imagine whether two four-dimensional spaces in a six-dimensional space can be continuously deformed to each other. By contrast, algebraic objects are quite rigid. Associating these rigid "invariants" to topological spaces proved the most fruitful method for distinguishing topological spaces and approaching the problem of classifying particular families of topological spaces.

Category theory was initially developed as a tool for abstracting the commonalities between these different homology theories and providing a unified framework for performing homological calculations. There are many different kinds of categories: categories of sets, categories of groups, categories of vector spaces, categories of topological spaces, and so on. The basic components of each such category are *objects* and *morphisms*. Intuitively, objects are the structures in question and morphisms are the appropriate mappings between these structures. For instance, in the category **Set**, the objects are sets and the morphisms are functions between sets; in the category **Top**, the objects are topological spaces and the morphisms are continuous functions between topological spaces.

Crucially, it is possible to define maps between categories (functors) and maps between functors (natural transformations). Using these tools, it is possible to transfer properties from one category to another, and hence results and constructions from one area of mathematics to another. As Allen Hatcher poetically notes, functors provide the "lanterns of algebraic topology" projecting algebraic images of topological spaces.[18]

10

Although not originally conceived in foundational terms, category theory abstracts many of the fundamental patterns and constructions appearing in mathematics. Furthermore, certain constructions that are intuitively clear in familiar categories like **Set**—for instance products, disjoint unions, and forming the collection of functions between two sets—can be generalized and shown to exist (or not) in other categories. As certain general mathematical constructions can be carried out in any category possessing an appropriate list of properties, one can organize vast mathematical conceptions from a simple axiomatic standpoint, providing something of an atlas of the world of mathematical constructions.

It was quickly observed that many logical concepts could be expressed in categorical terms. In the present context, the most important one is the notion of a *model*. In logic, a model provides an interpretation of some collection of syntactic statements, a structure on which all of the statements hold. For instance, one might have a collection of algebraic equations—say, the axioms of a group. The collection of all statements that follow from these axioms would be an "algebraic theory," namely the "theory of groups" (often written $Th(G)$). Note $Th(G)$ is a purely syntactic collection, but at the same time provides all statements derivable from the axioms of group theory. Of course, these statements will all hold on any specific group of interest. Intuitively, a model picks out a specific group satisfying all the statements. It assigns to the variables in each equation elements of the group under consideration.

In his 1963 thesis, *Functorial Semantics of Algebraic Theories*, F.W. Lawvere saw that these "algebraic theories" could be axiomatized categorically, in terms of categories having "finite products." Given this context, he showed that a model of an algebraic theory was a *functor* from a category with finite products to the category **Set** (which preserves finite products). Hence a model can be viewed as an *arrow* or *map* between categories; a model provides the "meaning" or "interpretation" of a collection of syntactic statements.

This notion that arrows represent meaning is a recurrent theme in Hennix's work. For instance, adopting a Wittgensteinian tone, she notes that "the world consists of arrows . . . each arrow is a fact—a mental fact."[19] As we shall discuss momentarily, she derives her notion of "algebraic aesthetics" from the algebraic semantics of Lawvere.

11

In classical mathematics, we have seen set-theoretic constructs used to form constant structures, to which "spatial" properties are added as an additional layer of structure. In this classical framework, continuous variation of space is modeled through the notion of a sheaf. A sheaf can be viewed as rule F, which assigns to each point x in a topological space X a set F_x consisting of the "germs" at x, typically taken as the functions defined in a neighborhood of x. The sets F_x for all x can then be "pasted" together to form a space projecting onto X. Viewed in this light, the sheaf F is a set F_x, which "varies" (with the point x) over the space X.[20] The base space X is called the *domain of variation*.

The collection of sheaves on a given topological space X form a category **Shv**(X). Sheaves are one of the most fundamental topological constructions. In the early 1960s, Alexander Grothendieck developed a generalization of this notion to settings far beyond sheaves of functions on topological spaces. These generalizations were categories called *toposes*, and he considered them to be "generalized spaces" or "spaces to do mathematics." Just as cohomology was originally developed to give algebraic invariants of a topological space, Grothendieck showed how cohomology could also be used to find algebraic invariants of a topos. He then used this technique to solve some of the most important problems in twentieth-century mathematics.

12

As early as 1963, Lawvere began to conceive of category theory as a viable alternative foundational position for mathematics. Instead of sets and membership, the new irreducible concepts could be objects, morphisms, functors, and natural transformations. In fact, since objects could in some sense be identified with identity morphisms, the notion of an arrow could be taken as the underlying primitive concept of mathematics, without reference to further set-theoretic constructs. As Colin McLarty astutely notes, in set theory the only structure preserved by functions is identity, whereas in category theory "arrows reveal structure" or "arrows are used to define structure."[21] Of course, this required translating set theory (or set theory and some part of first-order logic) into the language of category theory, and in 1964 Lawvere published his initial attempt, *The Elementary Theory of the Category of Sets.*

In 1966, Lawvere encountered Grothendieck's notion of a topos, which he initially conceived as a possible framework for developing a synthetic version of differential geometry. Between 1969 and 1971, Lawvere and Tierney axiomatized the notion of an (elementary) topos as a category with suitable "limits," "exponential objects," and "subobject classifiers." In fact, this axiomatization closely resembles Lawvere's categorical axiomatization of sets from a few years earlier. Each axiom of a topos also holds as an axiom of set theory. However, a few statements typically taken to hold in classical set theory (for instance the axiom of choice and a "well-pointedness" axiom) are absent from the axioms of a topos. Both of these excluded axioms are incompatible with continuous variation, and in fact fail on many categories of sheaves.

In this way, a topos permits a wider and more general conception of space than the standard set-theoretic perspective. Furthermore, it allows for a range of possible spatial structures for doing mathematics that are continuously variable in an intrinsic sense, and not reducible to discrete terms.

Lawvere envisioned his categorical approach explicitly with such notions in mind, and with avowedly philosophical motivations. He writes:

> The (elementary) theory of topoi . . . is a basis for the study of continuously variable structures, as classical set theory is a basis for the study of constant structures. . . . As Engels remarked in the period when set theory and the arithmetization of geometry did not yet dominate mathematical thinking, the introduction of the advance from constant quantities to variable quantities is a mathematical expression of the advance from metaphysics to dialectics, but many mathematicians continued to work in a metaphysical way with methods which had been obtained dialectically (*Anti-Duhring*, in the section on Quantity and Quality). . . . Every notion of constancy is relative, being derived perceptually or conceptually as a limiting case of variation and the undisputed value of such notions in clarifying variation is always limited by that origin.[22]

As we shall see, this notion of constancy as the limiting case of variation provides a significant undercurrent to Hennix's writings, as well as to Young's notion of tuning as a function of time and Flynt's concept art.

As a final piece of mathematical background, one axiom of a topos bears special consideration: the existence of a "subobject classifier," often notated Ω. This can be used to describe the subobjects of an object in a category, and in some sense captures the internal logic of a topos. For instance, consider the example of a category of sheaves of sets on a topological space X. For any open set U of X, $\Omega(U)$ describes the set of all open subsets of U.

This is an important example. As mentioned earlier, for any topological space X one can form an algebra of open sets of X. It is easy to see that this algebra can be represented in terms of propositional logic. A propositional variable p can be associated to some particular open subset U of X (in other words, in the relevant topological semantics, propositional letters are assigned to open subsets of X). However, $\neg p$, the complement of p in X, is not in general an open set, and hence may not be in the relevant algebra. Instead, $\neg p$ should refer to the union of all open sets contained in the complement of p in X (the union of any collection of open sets is always open). In particular, this entails that $\neg\neg p$ does not imply p: the complement of the complement of an open set need not equal the open set with which one started. This is logically equivalent to saying it is not the case that "p or $\neg p$" holds. Hence in general the law of excluded middle fails for the internal logic of a topos. It may hold on some toposes, and not hold on others.

Recall this treatment of negation (and the law of excluded middle) also underlies the logic behind Brouwer's intuitionistic continuum. For instance, there are many real numbers (so-called weak counterexamples) where $x < 0$ or $x \geq 0$ does not hold. In both cases, it is the dynamic nature of the structures in question that precludes certain either-or statements from holding.

In general, the internal logic of a topos is intuitionistic. The relevant algebra is called a Heyting algebra (named after Brouwer's student who developed intuitionistic logic), or sometimes a Brouwerian lattice. As Lawvere notes:

> The internal logic of a topos is always concentrated in a Heyting algebra object. If this object happens to be Boolean [i.e. a Heyting algebra with classical negation], then the variation of the sets is (constant or) random in the sense that for every part b of the domain of variation the topos splits as a full product $\mathscr{C} = \mathscr{C}/b \times \mathscr{C}/b'$, i.e., any motion over b and any motion over the complementary part can be combined into a total motion admitted by $\mathscr{C}$, whereas for most topoi there is a continuity condition at the boundary of b; this is of course analogous to the contrast between continuous and measurable variable quantities.[23]

If the subject classifier is a Boolean algebra, it has the typical separability properties of the classical continuum. However, in a more general Heyting algebra that is not Boolean, stronger coherence properties may apply, as occurs in Brouwer's intuitionistic continuum.

14

In 1971, Hennix arrived at UC Berkeley as an exchange student. As detailed earlier, she began work on her *Infinitary Compositions*, which stretch through much of the 1970s. In the summer of 1973, she wrote a proposal to teach a course on this topic at Mills College. The proposal still exists, although the course was never offered.

Her early pieces such as *Forcing*, *Fixed Points*, and $\Box_\kappa$ are written in the language of set theory and concern Boolean-valued models, large cardinals, and other topics. As mentioned earlier, Hennix found Solovay's work on these topics influential. Indeed, parts of *Forcing* appear to be derived from Section 1.1 of Solovay's *A Model of Set Theory in which Every Set of Reals Is Lebesgue Measurable*, or perhaps course notes on this topic. In comparison with her compositions from a few years later, Hennix colors within the lines of set-theoretic formalism in these early pieces, with none of the extra-mathematical interpolations and excursions that characterize her later work. In one sense, these pieces are more restrictive than her later efforts, in which mathematical terminology is blended and given additional treatment layers. For instance, a few years later, she theorized encountering her infinitary composition $\Box^N$ as follows:

> At the moment the Creative Subject enters a frame for a topos $\Box^N$ as defined by the interior of E', the time manifold ^{N} is supposed to already contain indefinitely many time elements which have not yet or never will be experienced by the Creative Subject. . . . [Later] at "transfinite stages" of experiencing $\Box^N$ the Creative Subject arrives at a stationary subset of the generated continuum of perceptions at which she retains complete faculative control of the continuum of time-elements which defines the transfinite time-object $_\kappa$, where κ denotes an ordinal of some large cardinal and not merely a large number as connoted by the symbol N.[24]

Although she employs a considerable amount of set-theoretic terminology, Hennix mixes together Brouwer's intuitionism, topos theory, topology, and acts of introspection. She combines a set-theoretic result concerning the existence of fixed points on any monotone increasing function from ordinals to ordinals, the experience of a "constant event" of a repeating composite waveform, and Brouwer's Creative Subject to describe a composition involving three sine tones. In essence, the Creative Subject encounters these rationally tuned sine tones, and the "topos" in question can be viewed as a state of consciousness or the acts of attention that the Creative Subject chooses to devote to this composite waveform. The constant event is the experience of the composite waveform repeating in time. We will describe this setup more precisely in a moment.

The sparsity of the composition relates to attaining maximal clarity regarding one's own acts of cognition. (Hennix often refers to Occam's razor or "the law of sufficient reason" when weighing the addition of elements or structural components to her compositions.) In this way, the sound waves function like notation or syntax for focusing one's attention, similar to the perceived length-to-width ratios in Flynt's *Illusions-Ratios*. We will discuss the relationship of sine tones to Brouwer's "empty two-ity" shortly.

Following Brouwer, Hennix believes that spare introspective acts, such as the act of doing mathematics, can lead to mental freedom from what she calls "the Mundane World," or what Brouwer called "the Sad World." For instance, Hennix notes that the "topos" under which one attends only to one's act of attention (i.e., Brouwer's empty two-ity) is where "the Creative Subject realizes a complete sense of (perceptual or cognitive) freedom."[25]

15

While Hennix's earliest infinitary compositions have a more cohesive quality internally, in the sense that they are located within a pure set-theoretic formalism, viewed externally they are in some ways even more cryptic. They are compositions in an abstract sense, as there is no obvious auditory content. The sole acoustical reference occurs in *Fixed Points*, where she provides a method of enumerating just intonation musical intervals in passing. The compositions offer little indication of their meaning to someone not enrolled in the topics courses in set theory Hennix attended, and even then shed little insight into how they might function as compositions in a traditional sense.

Like her writings, these compositions divulge little context of the thought process surrounding their creation. They seem to exist purely as internal acts in the mind of the observer, in which the state of mind or attitude generated by the piece (regardless of content) becomes a part of the composition itself. As we will discuss later, there is a certain Cageian aspect to this reasoning. In slightly more general terms, however, these pieces also reflect on the nature of musical intervals in terms of different conceptions of mathematics, and on what constitutes a continuum of experiences (for instance, in *Forcing*, she attempts to construct a "continuum of [musical] intervals" as fixed points of an ordinal function). From this perspective, the pieces provide a nascent picture of themes that will develop in her work.

Musical intervals are commonly represented as frequency ratios, and hence may be "arithmetized" in a fashion similar to the classical continuum. An octave has the frequency ratio 2/1 and so can be associated with the real number 2/1= 2; a fifth can be understood as the frequency ratio 3/2 = 1.5, a fourth 4/3 = 1.33 . . . , and so on. Furthermore, some common intervals are classically represented as irrational numbers: for instance, a tritone has the frequency ratio $\sqrt{2}/1$. Following this reasoning in a suitably abstract fashion, any point on the classical continuum, or on the continuum (1, 2), can be identified with a musical interval to form one model of a "continuum of musical intervals."

However, individual points on the classical continuum are located with infinite precision on a line, which is not replicable in any acoustical environment. Another difficulty with this correspondence is that there are an uncountable infinity of distinct points on any line segment. Thus uncountably many musical intervals "exist," yet there is no way to refer to them or to know if they have been performed.

While a familiar mathematical quandary, this question of abstract "existence" as opposed to realizability became a theme central to conceptual works coming out of the 1960s avant-garde, in particular the word scores surrounding the publication of *An Anthology*. For instance, Young's *Composition 1960 #15* reads: "This piece is little whirlpools out in the middle of the ocean." This is an unusual piece for Young; even the most conceptual of his word pieces are in general performable and moreover realized in a highly constructive manner. However, this specific piece provides an archetypal nonconstructive statement. It asserts the existence of an object without providing a means of locating it.

Flynt's response to Young's *Composition 1960 #10* (in which a line is printed on a note card) is an even more explicit commentary on this topic. The 1961 "score" contains a line drawn on a sheet of paper together with the statement *Each Point on this Line Is a Composition*. Flynt's piece asserts the "existence" of an uncountable number of compositions (or at least some continuum of compositions), almost all of which in a classical setting are both unrealizable and unidentifiable. Flynt's 1961 *Work Such That No One Knows What's Going On* operates along similar lines, and in fact one could view *Each Point* as a model of it.

17

The notion of modeling a continuum of experiences (or a continuum of intervals) is a theme common to both Hennix's writing and Flynt's concept art, and as we shall see shortly, is implicit in Young's notion of tuning as a function of time. Through the influence of Young, Hennix was predominantly concerned with just intonation intervals, i.e., intervals represented by ratios of whole numbers. As rational numbers can be individually enumerated and constructed in an intuitive manner, they would seem less in need of any conceptual revision or clarification than the classical continuum. Nevertheless, even under Brouwer's construction they have somewhat different properties, in the sense that they remain a dynamic, infinitely proceeding structure rather than a completed set.

Under different foundational assumptions, for instance Esenin-Volpin's ultrafinitism, even the natural number series loses its uniqueness and rigidity. This is a relevant example, as Hennix met Esenin-Volpin through the composer Maryanne Amacher in the early 1970s, and eventually became his student and collaborator. Esenin-Volpin's conception of distinct natural number series became a significant component of Hennix's work from the mid-1970s on. Many of her compositions, writings, and visual art reference Esenin-Volpin's concept of "short infinitary processes," and she considers different conceptions of a continuum of intervals once the natural number series has lost its uniqueness.

Interestingly, this notion of nonunique number series also arises in category theory. Any category having a "terminal object" has a so-called natural number object, which behaves similarly to the natural numbers. However, under sufficiently weak assumptions, these natural number objects may fail to be unique (i.e., there may be nonisomorphic natural number objects in the relevant category). In *The Yellow Book*, Hennix attempts to connect these categorical natural number objects to some of Wittgenstein's remarks on the foundations of mathematics. She notes, "The logical object which corresponds to the natural numbers, $\mathbb{N}$, in the *Tractatus*, can be interpreted as a continuous variable set of natural numbers $\mathbb{N}$ related to a topos construction satisfying the usual Kock-Lawvere axioms where the difference between two external natural number objects $\mathbb{N}_1$, $\mathbb{N}_2$ are interpreted as different stages of the construction of the internal natural number object $\mathbb{N}_0$."[26]

A topos under the standard Lawvere axioms has up to isomorphism a unique natural number object, so that must be the internal natural number object to which she refers. She claims that different "external" natural number objects can be interpreted as parts of a construction series for this "continuously variable set of natural numbers." She does not make the exact correspondence to Wittgenstein's philosophy of mathematics explicit; however, one can see her reflecting on the effects of removing the rigidity of the natural number series. Hennix later used this idea to theorize some of her compositions and visual art, in particular her *Short Infinitary Process* paintings.

From a more general compositional perspective, one can begin to detect a pattern in her work. If numbers model music, how does some foundational shift—for instance from classical to intuitionistic mathematics, or from constant structures to variable ones—influence the correspondence with musical intervals?

18

The most thorough analysis of this sort comes through Hennix's engagement with Brouwer and her theorization of the music of Young. As detailed in my essay "Minimalism and Foundations," facets of Young's music bear a strong resemblance to Brouwer's time- and subject-dependent version of mathematics.

For instance, several of his word scores in *Compositions 1960* suggest potentially infinite constructions of the basic elements of music, geometry, and arithmetic, carried out through successive acts of introspection. *Arabic Numeral (Any Integer) to H.F.* (1960) describes a loud piano cluster to be repeated some given (but unspecified) number of times with as little variation as possible. This suggests a potentially infinite construction of a natural number series through the repetition of introspective acts, similar in spirit to Brouwer's conception of a number series being a "repetition of 'thing in time' and 'thing again.'"[27]

Recall that for Brouwer, the basic "two-ity" is a sensation stored in memory as its passing away, being replaced by another sensation. In the special case of the empty "two-ity," a self-same sensation of an empty cognitive state is replaced by another. Through its observation of self-similar events, Young's *Arabic Numeral* resembles Brouwer's construction of a two-ity, a three-ity, and so on. However, in this auditory version, the self-same act of introspection is based on the notion of listening to musical intervals. Later, Young and Hennix would specialize this to the case of rationally tuned intervals and their repeating composite waveforms. Along similar lines, Young's *Composition 1960 #7* gives a potentially infinite construction of one of the basic elements of music: it notates a perfect fifth to be held "for a long time."

Perhaps most suggestive, *Composition 1960 #10 to Bob Morris* provides the instructions to "draw a straight line and follow it." While this could be taken as a conceptual exercise, it reflects Young's compositional process in that the piece is not only performed but carried out in a highly constructive manner. In the initial 1961 performances at Harvard (organized by Flynt, and realized by Young and Robert Morris) and at Yoko Ono's loft, Young and Morris determined a sight, along with a point in the vicinity of where the line should end. Every few feet a plumb bob was aligned visually with the sight, with Young providing verbal directions on how to adjust the plumb. They made chalk markings on the floor, and later connected all markings with a yardstick. As in the process of tuning, the line was only built up over time through successive perceptual adjustment. While an elementary form was investigated, it was not treated as an external reality referred to by performance, but rather something constructed in time through the subject's perspective.

In 1962, Young encountered just intonation, and from this point on the audible structure of the harmonic series became a central principle of organization in his music. The addition of tuning suggests an important theoretical refinement to Young's approach: not only are structures of music potentially infinite, but so are the elements themselves. Comparing tuning to the astronomical observation of planets in orbit, he notes, "Tuning is a function of time. Since tuning an interval establishes the relationship of two frequencies in time, the degree of precision is proportional to the duration of the analysis, i.e. to the duration of tuning. Therefore, it is necessary to sustain the intervals for longer periods if higher standards of precision are to be achieved."[28]

Young goes on to argue that the accuracy of a tuned interval corresponds to the number of observed cycles of its periodic composite waveform. Instead of completed, "arithmetized" points on a continuum, intervals are given as constructions developing in time based on an idealized observer's attention. Hence they resemble Brouwer's choice sequences. Rather than idealized objects placed with infinite precision on a line, they are dynamic constructions requiring longer acts of attention for higher standards of precision. As Young notes, even in the simplest case of unison, it could take hundreds of years or more for one complete cycle of a composite waveform to occur (for frequencies sufficiently close together). As in Brouwer's choice sequences, there is always a slight halo surrounding the status of an interval, given its essentially dynamic nature.

Furthermore, for irrational intervals there will never be even one complete cycle of a composite waveform. Young interprets this to mean that irrational intervals such as tritones are untunable. (Flynt constructed his *Tritone Monochord* [1987] as an objection to this claim.) Of course, under a classical interpretation, given any finite amount of time it is always possible to determine an explicit irrational number close enough to any given fraction such that one complete cycle of a composite waveform would take longer than the specified amount of time. Distinguishing between the two intervals would then take longer than that specified period of time, so it is not strictly possible to tell whether a rational or irrational interval has been performed. Indeed, this may have been part of the reasoning behind Hennix's search for a new continuum of musical intervals. On the other hand, this statement only concerns Young's theory of tuning. In practice, the just intonation intervals he engaged with were bounded by certain auditory principles, so with effort one could work out a suitably restricted version of the theory having the intended consequences.

Young's principle is demonstrated in many of his compositions from the 1960s on. The *Drift Studies* of the late 1960s and '70s involve rationally tuned sine tones gradually going in and out of phase over periods of observation. (It is worth noting that although Young refers to "tuning as a function of time" as one of his key theoretical constructs, the theory may have stemmed in part from technological limitations of the time, in particular the instability of commercially available oscillators in the 1960s.) Similarly, his *Four Dreams of China* of 1962 (and many later versions) involve sustained voicings of a single four-note chord and its subsets. One can see Young's principle suggesting these pieces as further developments of potentially infinite constructions of musical intervals, rather than suspensions of completed forms.

19

Hennix met Young in New York in 1969, and in the same year Young commissioned her to realize one of his *Drift Studies* at the EMS studio in Stockholm. (EMS had recently acquired several phase-locked oscillators that allowed for much more stable versions of the *Studies*.) She also intended to realize a sine tone version of Young's *Trio for Strings* (1958), although this project was never completed. Shortly after meeting Young, Hennix set out to write computer music for rationally tuned sine tones. Her method of composition is closely modeled after Young's mid-1960s works, but she theorizes her approach more broadly, in terms of intuitionism, category theory, and set theory.

As noted above, Brouwer's thinking overlaps with several tenets of the 1960s avant-garde. Most prominently, his mathematical systems are composed of acts of attention: a first sensation stored in memory at its passing away, being replaced by another sensation. This conception fits nicely within the framework of the post-Cage avant-garde, in which a composition is treated as a construction of mental states or located in the attitudes of the observer rather than existing as a completed, external state of affairs.

Through the influence of Cage, this notion became broadly prevalent. Although any number of examples could be chosen, the idea is perhaps most concisely expressed in an early word piece by the mathematician, composer, and Young collaborator Dennis Johnson: *LISTEN* (1960). It is easy to trace the influence of this idea on Hennix's work via Cage, Flynt, and Young. Concept art sought a formalization of the construction of mental states common to works of the 1960s avant-garde. Flynt looked at the ambient aesthetic doctrine surrounding Cage's work (most obviously 4′33″ [1952] but also his concept of indeterminacy) and sought to "apply new music to metamathematics."[29] Similarly, Young's focus on the audible structure of the harmonic series and other acoustical properties of sound emphasized the act of listening in the compositional process.

20

Given Cage's influence on Young and Flynt, it is worth considering the kinds of mental states his compositions were supposed to bring about. While some think Cage's use of chance operations was intended to generate novel compositions free from the obstructions of ego, it is perhaps most apt to characterize the newness or novelty of these compositions in terms of experiential states or mental attitudes, rather than externally defined states of affairs.

Echoing Lawvere's remarks about constancy being the limiting case of variation, Cage notes, "If we think that things are being repeated, it is generally because we don't pay attention to all the details. But if we pay attention as though looking through a microscope to all the details, we see that there is no such thing as repetition."[30] Likewise, he writes, "New music: new listening. Not an attempt to understand something that is being said, for if something were being said, the sounds would be given the shape of words. Just an attention to the activity of sounds."[31]

In fact, when Cage speaks of newness it is typically in relation to a certain kind of listening, which he refers to as a "purposeful purposelessness or a purposeless play. This play is an affirmation of life—not an attempt to bring order out of chaos."[32] Of course, this disposition Cage describes is not itself new; the notion of a "purposefully purposeless" act of attention comes from Kant's theory of aesthetic judgment, although Cage himself likely encountered it via Dewey. Compare Cage's aforementioned statements to Kant's *Critique of Judgment*: "*Beauty* is an object's form of *purposiveness* insofar as it is perceived in the object *without the presentation of a purpose.*"[33] Kant writes,

> The basis that determines a judgment of taste, can be nothing but the subjective purposiveness in the presentation of an object, without any purpose (whether objective or subjective), and hence the mere form of purposiveness, insofar as we are conscious of it, in the presentation by which an object is given us. . . . An aesthetic judgment refers to the presentation of the imagination, by which an object is given, solely to the subject; it brings to our notice no characteristic of the object, but only the purposive form in the way the presentational powers are determined in their engagement with the object.[34]

Like works of Cage and Young, Kant's judgment of aesthetic beauty does not reflect on the object presented but rather "the purposive form in the way the presentational powers are determined in their engagement with the object." However, for Kant, judgments of beauty are not formed with respect to art (which he viewed as a secondary phenomenon) but rather to experiences of nature. For instance, in observing a flower one might enjoy perceiving an order to its structure. Yet upon further reflection we notice that this order exists nowhere outside of us, and what is actually experienced is "the presentation by which an object is given to us." It is this mental state that underlies Cage's compositional attitude, and the particular notion of "freeness" that Kant describes when encountering beauty will become important in a moment.

Via Cage, one can see a relation of concept art to more classical theories. Flynt's "purely aesthetic proofs" explore the mind's presentational powers, with the "syntax" of the language being a reflection on how objects are presented to us. Through Flynt, this formalization of the mind's faculative powers becomes significant to Hennix.

21

One significant component of the cognitive freedom underlying Brouwer's approach to mathematics was an experience of a play of the mind's presentational powers. As noted above, this was also a central feature of Cage's compositions, Young's sustained compositions, and Flynt's concept art.

Filtering this notion through the work of Lawvere, Hennix developed the idea of a topos representing a state of consciousness or modality of thought. It was noted earlier that "algebraic theories" can be represented by categories with finite limits, and that "algebraic theories of proofs" (i.e., λ-calculus) could be represented as Cartesian closed categories. Furthermore, these algebraic theories are purely syntactic objects, either presented or generated freely as strings of symbols. Following Flynt, Hennix associates such syntactic objects with mental acts or some display of the mind's presentational powers.

Just as there are different modes in music expressing different emotional states, and in math different toposes corresponding to different "spaces to do mathematics," Hennix links different toposes to alternative "places of thought" or states of awareness from which to perceive the world. She frequently refers to the Creative Subject as a free stream of sensations, and to a topos as some more structured state of awareness. One can see this view as an extension of Cage's and Flynt's work, permitting additional, more structured modalities of awareness by adding further specifications, analogous to different modes in music.

However, unlike Flynt, Hennix's approach contains an explicitly stated semantic component. In category theory, a model of an algebraic theory is a functor from the appropriate category to the category **Set** (preserving some properties). While as a category, an "algebraic theory" (of which a topos is one particular instance) is a something like a "free" or "presented" object, a model for this category picks out a specific structure satisfying the conditions, or assigns a meaning to a collection of syntactic statements. Functors (or arrows) determine structure, a concept Hennix draws upon extensively in her work.

Following Lawvere, Hennix interprets a model as an arrow between categories. However, under her nonstandard interpretation of a topos, a functor realizes a state of consciousness or is an interpretation of mental states. Modifying the beginning of the *Tractatus*, she writes, "The world is formed of arrows. Each arrow is a fact—a mental fact. A mutual fact. . . . Philosophical problems can be analyzed as stacks of arrows."[35]

In response to Flynt's concept art, in the mid-1970s Hennix developed her notion of Epistemic Art. These works involve "arrows" or interpretations out of spare syntactically presented objects. For instance, in the series *Algebras w/ Domains* (1973–91), red, blue, black, and white squares are painted on a grid, along with "equations" such as "a blue box followed by a white box equals a white box followed by a blue box" (she gives other, more complicated identities as well). The paintings are intended to specify "equational" or "algebraic theories." One can see them as syntactic presentations of algebraic objects given as "free" objects (which are just strings of symbols, or in this case strings of colored boxes) modulo some relations that allow for "word reduction." (Of course it seems this string reduction will have to take place mentally.)

These syntactic objects can be interpreted on any structure meeting the requirements. In this case there is no "standard model," and possibilities abound. Just as one interprets the theory of groups, or alternatively some string of symbols in a free group modulo some relations, on a specific group, one "interprets" the free or presented objects in the paintings on structures in the world. As we will see momentarily, her spare sine tone compositions function similarly, as "equational theories."

Given the abundance of interpretations permitted, Hennix felt she had shifted the perspective on the "syntax rich, semantics poor" works of Flynt and Cage. Rather than a purely syntactic presentation of the mind's presentational powers, Hennix added a substantive semantical component. Modifying Lawvere's popular mantra from category theory, she contends that "our slogan has now become 'Syntax poor, Semantics rich.'"[36] (Although one can see the syntax in both cases as fairly similar, Hennix formalizes her syntactic theory in the language of category theory and notes that models are appropriate arrows, as Lawvere did in his thesis.)

22

In 1976, there were two major presentations of Hennix's work at Moderna Museet in Stockholm. The first was *Brouwer's Lattice*, a weeklong concert series featuring several of her compositions, as well as a sound installation and pieces by Young and Terry Jennings. The second, a museum exhibition entitled *Toposes and Adjoints* (subtitled "A survey of abstract thought from Cantor to Lawvere"), featured texts, sculpture, sound installation, paintings, computer monitor displays, and incense, among other elements. A selection of titles gives a good indication of Hennix's interests: *Model for Set Theory, Open Point Set of Measure Zero, Brouwer's Bar (Answering a question of Walter De Maria using Brouwer's Bar Theorem), Short Infinitary Process, The Least Non-Measurable Cardinal, The Least Measurable Cardinal,* and *Topos #1–14,* among others. As in her writing, a loose and fully kaleidoscopic vision is evident, reminiscent of other 1960s-style world-building art.

Let us conclude with some remarks about these exhibitions. First, recall that a Brouwerian lattice is another term for a Heyting algebra, which expresses the internal logic of a topos. These toposes allow for generalized spaces of continuously variable structures, rather than just the special case of constant ones. Following Lawvere, Hennix chose this title as an expression of a move away from treating compositions as constant structures, and toward a more dynamic or synthetic approach, reflected in Young's "algorithmic" compositions of the late 1960s and his notion of tuning as a function of time. In the notes to *Brouwer's Lattice,* Hennix observes, "We owe the inauguration of this tradition," which she terms Brouwer's doctrine, "to La Monte Young and his fabulous New York City group of the 1960's [*sic*]. . . . [One aspect of this approach is] a focus on evolving frames of musical structures rather than trying to obtain completeness. In this context completeness is misconceived."[37]

Hennix goes on to describe the theory behind her own computer-generated algorithmic infinitary compositions, expressed in terms of Brouwer's notion of a choice sequence and Young's algorithmic compositions. She writes, "The computer adds to the technical precision. You are never limited by physical exhaustion and there are no obstacles for proceeding with infinitely long spreads of musical events locked together by some appropriate algorithm that generates each new step on the basis of the preceding ones."[38] Unfortunately, she never realized these choice sequence compositions, in part due to technical limitations of the time.

More generally, Hennix connects a Brouwerian lattice with a synthetic notion of space. In *The Yellow Book,* she writes:

> Given a topos $\mathscr{E}$ and an object X in $\mathscr{E}$ . . . and any point $p \in$ X the points well separated [i.e. of positive distance] from p are $x \in \neg\{p\}$ while the points not well separated from p are the points $x \in \neg\neg\{p\}$ i.e. those points that are infinitesimally close to p. Since intuitionistically X = $\neg\neg\{p\} \cup \neg\{p\}$ is false in general there must be some space unaccounted for in X. Remarkably this space is a "grey

area" (a "no man's land") because it has no explicit description in the topos.[39]

Here Hennix attempts to describe the role of a subobject classifier in a topos. Adopting a standard interpretation, she describes a topos as a kind of space, and a general Heyting algebra as a space with "grey areas," where the union of a set and its complement may not comprise the entire space. While this example could be more precisely articulated in terms of an algebra of open sets, she connects the notion of "space" in a topos to Brouwer's construction of the continuum, in which there is always a halo of "blurriness" as to where choice sequences are located. For instance, we have seen there are choice sequences x where $x \leq 0$ or $x > 0$ is false. It is these "grey areas" that give Brouwer's continuum its strong coherence properties, which Hennix chooses to express in terms of a non-Boolean Heyting algebra.

23

Several other pieces in *Brouwer's Lattice* and *Toposes and Adjoints* are theorized in terms of Lawvere. For instance, the floor painting *Open Point Set of Measure Zero* (1976) contains black splotches on the ground in a random-looking pattern. Hennix writes that "different phases of the continuously variable sets are shown as a continuously variable floor painting. . . . Thus the floor painting serves simultaneously as both a picture and as a model for a finite set of topoi constructed in the style of F.W. Lawvere."[40] Again, her use of Lawvere suggests a synthetic approach toward composition and visual art, expressed through cognitive structures.

The sound installation in *Brouwer's Lattice* is Hennix's infinitary composition from 1976, *The N-Times Repeated Constant Event* (sometimes written $\Box^N$). Following Young's sine tone pieces of the late 1960s (and referencing his *Arabic Numeral [Any Integer] to H.F.*), the constant event is understood to be one complete cycle of a composite waveform of three rationally tuned sine tones.

While in *Arabic Numeral* integers are linked to Brouwer's notion of time passing through direct experiential means, Hennix follows Brouwer in articulating a more primordial or idealized account of this procedure. Paraphrasing Brouwer's "empty two-ity," Hennix writes that the moment the subject comes to intuit the fundamental process of a waveform repeating in time "corresponds to a point in her life-world where a moment of life falls apart with one part retained as an image and stored by memory while the other part is retained as a continuum of new perceptions."[41]

Instead of realizing a two-ity, a three-ity, etc., through a repetition of self-similar piano clusters, Hennix interprets Brouwer's basic mathematical system in terms of an observation of complete cycles of composite waveforms. Here Brouwer's primordial intuition of time passing is directly tied to Young's notion of tuning as function of time. In addition, Brouwer's "empty two-ity" is now expressed in terms of the experience of "empty sounds," in the sense that sine tones are waveforms divested of familiar qualities such as overtones.

24

Later in The *Yellow Book*, Hennix theorizes *The N-Times Repeated Constant Event* in slightly different terms. Since her theorization of the piece serves as a nice summary of the topics discussed in this essay, it seems a good way to conclude. In the following passage, she articulates three rationally tuned sine tones as presenting an equational theory (a collection of sine tones being representable in terms of trigonometric equations):

> The triad $\langle x, y, z \rangle$ defines a class of equations $E(x, y, z)$ which in turn determines a certain set of *algebraic theories*, $Th(\mathbf{E})$ among which triples $\langle X, Y, Z \rangle$ and co-triples $\langle X^{co}, Y^{co}, Z^{co} \rangle$ have played an important role in my early formulations of an ALGEBRAIC AESTHETICS.

> For each Creative Subject Σ who attends the space $E'_{\square}$ in which the composition $\square^N$ is developing freely, there exists a (subjective) time interval, τ^*, during which the Creative Subject experiences a *constant event* $\square^*_{\langle x,y,z \rangle}$ as a *subobject* of the $E'_{\square}$-universally infinitely proceeding sequences of geometrically congruent composite waveforms $\square^{\tau}_{\langle x,y,z \rangle}$.

> In general the constancy of the subjective event $\square^*_{\langle x,y,z \rangle}$ can be experienced as also fluctuating so that the subobject $\square^*_{\langle x,y,z \rangle}$ of $E'_{\square}$ (created by a Creative Subject's attendance) as a submanifold $\square^{N*}$ of $E'_{\square}$ defines a *continuously variable or discretely variable* set (of (fluctuating) experiences of a temporal constancy of time).[42]

This passage gives a good indication of both Hennix's aims and the density of her texts, which often read like a private language or direct transcription of internal thoughts. While some notation is not defined and different concepts are blended together, one can sense a theory in the process of formation.

Like her *Algebra w/ Domains* paintings, the three sine tones are interpreted as spare syntactic objects yielding an "equational theory." This syntactic theory of sound waves is "freely developing," and can be observed by a Creative Subject in some given space $E'_{\square}$. Of course, since the sine tones are tuned in simple whole number ratios, the composite waveforms will repeat in self-same patterns. However, the specific standing wave pattern in the environment

$E'_\square$ is determined by the space in which the composition is located, the speaker arrangement, humidity, etc. Hence the composition is not treated as a completed object but dependent on the specific surrounding environment.

Similarly, regardless of environmental complexity, the composition depends on an interpretation of this freely developing syntax. Hennix notes the Creative Subject may internally experience the composition as a subjective time interval in terms of its repeated waveforms. In practice this would occur very fast, but the observer's experience may exist on a different timescale than the waveforms themselves. As noted earlier, this experience can be described as a "point in her life-world where a moment of life falls apart with one part retained as an image and stored by memory while the other part is retained as a continuum of new perceptions," or, in Brouwer's terminology, an experience of an empty two-ity. Again, Hennix associates the experience of listening to sine tones with Brouwer's spare cognitive act of constructing positive integers. Indeed, sine tones lack harmonics (and hence timbre), making them perceptually very difficult to locate. This lends them a feeling of sounds divested of familiar qualities.

However, as Hennix notes, this "repetition of thing in time and thing again" is a synthetic experience, which she ties to Lawvere's philosophical remarks on the foundations of mathematics. She suggests we view $E'_\square$ as a "topos" created by the subjects' acts of attention, and the subjective experience of $\square^*_{\langle x,y,z \rangle}$ as a subobject of an object in this category. The internal logic of this topos is intuitionistic, i.e., it is described by a Brouwerian lattice that is not Boolean. Hence, she views the objects in this category as continuously variable structures rather than constant ones. In this setting, she regards the "natural number object" of this category (as described above) as continuously varying or fluctuating experiences of the temporal constancy of time. In the composition, one directly experiences Lawvere's remark about constancy being the limiting case of variation.

1   Catherine Christer Hennix, excerpts from *Parmenides and Intensional Logics* (unpublished manuscript, 1979), 2.

2   Henry Flynt, "Concept Art," in *An Anthology of Chance Operations*, ed. La Monte Young (New York: La Monte Young and Jackson Mac Low, 1963), 28. Alternatively, one might regard the syntax as being the model in *Illusions-Ratios,* or this distinction as being blurred.

3   Henry Flynt, "The Crystallization of Concept Art in 1961," 1994. Henry Flynt, "Philosophy," http://www.henryflynt.org/meta_tech/crystal.html.

4   Rudolf Carnap, *The Logical Syntax of Language* (New York: Routledge, 2007), 52.

5   Catherine Christer Hennix, "Notes on Intuitionistic Modal Music," in *Poësy Matters and Other Matters* vol. 2, *Other Matters* (Brooklyn: Blank Forms Editions, 2019), 39.

6   Gottfried Leibniz, "On Freedom," in *Philosophical Essays*, trans. Roger Ariew and Daniel Garber (Indianapolis: Hackett, 1989), 264.

7   Colin McLarty, "The Uses and Abuses of the History of Topos Theory," *The British Journal for the Philosophy of Science* 41, no. 3 (September 1990): 353.

8   See McLarty, "Uses and Abuses," for an interesting discussion, in particular on the importance of the development of arrow notation for representing functions.

9   L. E. J. Brouwer, "Intuitionism and Formalism," *Bulletin of the American Mathematical Society* 20, no. 2 (1913): 86.

10 David Hilbert, "Die Grundlagen der Mathematik," Abhandlungen aus dem Seminar der Hamburgischen Universität 6 (December 1928): 65–85. English translation in Jean van Heijenoort, ed., *From Frege to Gödel: A Source Book in Mathematical Logic, 1897–1931* (Cambridge, MA: Harvard University Press, 1967), 464–479.

11 Spencer Gerhardt, "Minimalism and Foundations," in *Simplicity: Ideals of Practice in Mathematics and the Arts*, eds. Roman Kossak and Philip Ording (New York: Springer, 2017), 223–34.

12 Dirk van Dalen, "The Return of the Flowing Continuum," *Intellectica: Revue de l'Association pour la Recherche Cognitive* 51, no. 1 (2009): 137.

13 L. E. J. Brouwer, *Collected Works 1: Philosophy and Foundations of Mathematics*, ed. A. Heyting (Amsterdam: North-Holland, 1975), 523.

14 van Dalen, "Return of the Flowing Continuum," 138.

15 Brouwer, "Intuitionism and Formalism," 86.

16 For a precise formulation and interesting discussion, see Rosalie Iemhoff, "Intuitionism in the Philosophy of Mathematics," in *Stanford Encyclopedia of Philosophy*, ed. Edward N. Zalta, Winter 2016, https://plato.stanford.edu/archives/win2016/entries/intuitionism/.

17   van Dalen, "Return of the Flowing Continuum," 143.

18   Allen Hatcher, *Algebraic Topology* (Cambridge, UK: Cambridge University Press, 2002), 21.

19   Christer Hennix, "The Yellow Book," in *Being = Space × Action*, ed. Charles Stein (Berkeley, CA: North Atlantic Books, 1988), 326.

20   For a more precise formulation, see Saunders Mac Lane and Ieke Moerdijk, *Sheaves in Geometry and Logic: A First Introduction to Topos Theory* (New York: Springer, 1992), 64.

21   McLarty, "Uses and Abuses," 365.

22   F. William Lawvere, "Continuously Variable Sets; Algebraic Geometry = Geometric Logic," in *Studies in Logic and the Foundations of Mathematics* 80 (1975): 135–36.

23   Lawvere, "Continuously Variable Sets," 136.

24   Hennix, "Yellow Book," 340.

25   Christer Hennix, *Brouwer's Lattice* (Stockholm: Moderna Museet, 1976), 13.

26   Hennix, "Yellow Book," 340.

27   Brouwer, "Intuitionism and Formalism," 53.

28   La Monte Young and Marian Zazeela, *Selected Writings* (Munich: Heiner Friedrich, 1969), 7.

29   Flynt, "Crystallization of Concept Art," 5.

30   Richard Kostelanetz and John Cage, "The Aesthetics of John Cage: A Composite Interview," *Kenyon Review* 9, no. 4 (1987): 110.

31   John Cage, *Silence* (Middletown, CT: Wesleyan University Press, 1961), 13.

32   Cage, *Silence*, 12.

33   Immanuel Kant, *Critique of Judgment*, trans. Werner S. Pluhar (Indianapolis: Hackett, 1987), 236.

34   Kant, *Critique of Judgment*, 221, 229.

35   Hennix, "Yellow Book," 326.

36   Hennix, *Brouwer's Lattice*, 4.

37   Hennix, *Brouwer's Lattice*, 13.

38   Hennix, *Brouwer's Lattice*, 13.

39   Hennix, "Yellow Book," 317.

40   Hennix, "Yellow Book," 317.

41   Hennix, "Yellow Book," 341.

42   Hennix, "Yellow Book," 343.

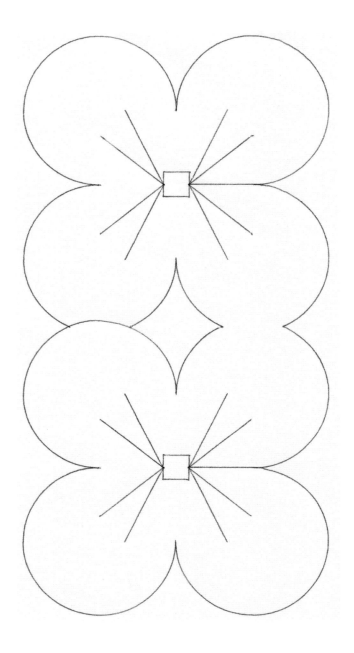

Matt Paweski
Study for Linked Cabinet, 2023
Ink on paper
Courtesy of the artist

A Construction Method for Modal Logics of Space
171

1   Introduction

Given the important role of spatial intuition in cognition, and the apparent unreliability of these intuitions, there is something natural about looking at spatial structures from an axiomatic standpoint. Indeed, it is not surprising that the first and best-known application of the axiomatic method was in providing a development of geometry.

In the past century, with the development of formal logic and the subsequent discovery that elementary number theory is not axiomatizable, it became both possible, and independently interesting, to examine spatial structures inside given logical formalizations. While first order logic has come to be accepted as the central framework for examining spatial structures axiomatically, from time to time other logics with spatial interpretations have been considered as well.

Recently, modal logics with spatial interpretations have proven a popular area of investigation. While this subject can be traced back to McKinsey and Tarksi's [10] work on boolean algebras with closure operators in the 1940s, within the past ten years a more general program of providing a modal analysis of space has emerged.

By and large, the techniques used in investigating modal logics of space have been model-theoretic in nature, involving the transfer of geometric or topological structure from the desired mathematical object to some Kripke frame. While this works well in cases where the relevant modal logics have nice Kripke frame characterizations, in other situations this way of proceeding can be quite difficult.

In this paper we examine a more syntactic approach for establishing completeness results for modal logics of space. The technique we use has the virtue of constructing the desired mathematical objects directly, rather than working indirectly through Kripke frames. This allows for a good deal of control over what the relevant models look like.

To start, we review the basic syntax and semantics for modal logics of space, and introduce our construction method. We next use the construction method to prove the strong completeness of the logic **S4** with respect to $\langle \mathbb{Q}, \tau \rangle$, extending a particular case of a general result of McKinsey and Tarski [10]. We then prove the strong completeness of the combined temporal and topological logic $\mathbf{Q}_{\Box \mathbf{FP}}$ for the rational line, extending a completeness result of Shehtman [14]. Finally, we use the construction method to establish that the temporal Since/Until logic $\mathbf{Q}_{SU}$ is strongly complete with respect to $\langle \mathbb{Q}, < \rangle$, extending a result of Burgess [6]. In the last chapter, we discuss the advantages and disadvantages of our construction method in comparison with the standard model-theoretic approach, using the Cantor space as a basis of comparison.

2 Modal Logics of Space

2.1 Syntax and Semantics

Although they would not have been thought of as such at the time, some of the first semantic completeness proofs for modal logic date back to McKinsey and Tarski's [10] work on the foundations of topology in the 1940s. These proofs pre-date the now standard Kripke semantics for modal logic, and in fact are spatial completeness proofs for the modal logic **S4** with a topological semantics.

Let $\mathcal{L}$ be the basic modal language, consisting of propositional variables $p, q, r..$, boolean connectives $\wedge, \vee, \neg, \rightarrow$, and unary operators $\square$ and $\diamond$. The formulas of $\mathcal{L}$ define the least set Γ containing all propositional variables which is closed under boolean connectives and the operators $\square$ and $\diamond$.

Definition 2.1. A modal logic **L** is the least set of formulas of $\mathcal{L}$ containing all propositional tautologies and the axioms of **L**, which is closed under substitution and proof rules:

$$\frac{\varphi, \varphi \rightarrow \psi}{\psi} \qquad \frac{\varphi}{\square\varphi}$$

Definition 2.2. A formula φ is said to be *provable* in **L** (written $\vdash_L \varphi$) if $\varphi \in$ **L**. A set Γ of formulas is said to be **L**-*consistent* if there is no finite set $\{\varphi_1, \varphi_2, ..., \varphi_n\} \subseteq \Gamma$ such that $\mathbf{L} \vdash \neg(\varphi_1 \wedge ... \wedge \varphi_n)$.

The central modal logic for our purposes is **S4**, which contains the axioms:

$\square\varphi \leftrightarrow \neg\diamond\neg\varphi$
$\square(\psi \rightarrow \varphi) \rightarrow (\square\psi \rightarrow \square\varphi)$
$\square\varphi \rightarrow \varphi$
$\square\varphi \rightarrow \square\square\varphi$

Recall that a *topological space* is a pair $\langle W, \tau \rangle$, where W is a nonempty set and τ is a collection of subsets of W satisfying the following properties:

- $\emptyset, W \in \tau$;
- if $U, V \in \tau$, then $U \cap V \in \tau$;
- if $\{U_i\}_{i \in I} \in \tau$, then $\bigcup_{i \in I} U_i \in \tau$;

Definition 2.3. A *topological model* $\langle W, \tau, v \rangle$ is a triple, where $\langle W, \tau \rangle$ is a topological space and v is a valuation assigning subsets of W to propositional letters.

Further recall that for any $U \subseteq W$,

- U is *open* if $U \in \tau$
- U is *closed* if $W - U \in \tau$
- The *interior* of U (written U°) is the union of all open sets contained in U
- The *closure* of U (written $\overline{U}$) is the intersection of all closed sets containing U
- $U^\circ = W - (\overline{W - U})$

In the original topological semantics for $\mathcal{L}$ proposed by McKinsey and Tarski, the truth of formulas was evaluated at the level of topological models. Given a valuation v assigning propositional letters to subsets of W, v extended to all formulas of $\mathcal{L}$ as follows:

- $v(p) \subseteq W$
- $v(\neg \varphi) = W - v(\varphi)$
- $v(\varphi \vee \psi) = v(\varphi) \cup v(\psi)$
- $v(\varphi \wedge \psi) = v(\varphi) \cap v(\psi)$
- $v(\varphi \rightarrow \psi) = (W - v(\varphi)) \cup v(\psi)$
- $v(\Box \varphi) = v(\varphi)^\circ$
- $v(\Diamond \varphi) = \overline{v(\varphi)}$

A formula φ was said to be true on a topological model $\langle W, \tau, v \rangle$ if $v(\varphi) = W$.

In addition to working with the notion of truth on the level of models, we will also make use of the notion of a formula φ being true at individual points. This fine grained approach mirrors the now prevalent Kripke semantics for modal logic, and will facilitate transferring techniques and results later on.

Given a topological model $M = \langle W, \tau, v \rangle$, we define a formula φ to be *true* at a point $w \in W$ by induction on the length of φ. From now on we will economize on boolean connectives, considering only $\neg$ and $\vee$, the additional operators being definable from these two alone.

- $w \models p$ iff $w \in v(p)$
- $w \models \psi \vee \chi$ iff $w \models \psi$ or $w \models \chi$
- $w \models \neg \psi$ iff not $w \models \psi$
- $w \models \Box \psi$ iff $\exists U \in \tau$ such that $w \in U$ and $\forall w' \in U, w' \models \psi$
- $w \models \Diamond \psi$ iff $\forall U \in \tau, w \in U$ implies that $\exists w' \in U$ such that $w' \models \psi$

Definition 2.4. We say φ is *true* on a topological model $\langle W, \tau, v \rangle$ (written $\langle W, \tau, v \rangle \models \varphi$), if $\langle W, \tau, v \rangle, w \models \varphi$ for every $w \in W$.

Notice that the above definition of truth is equivalent to Tarski's formulation, since $\langle W, \tau, v \rangle, w \models \varphi$ for every $w \in W$ if and only if $v(\varphi) = W$. Finally, in addition to truth on a topological model, we will also make use of the notion of validity on a topological space.

Definition 2.5. If φ is true on $\langle W, \tau, \nu \rangle$ for *any* valuation ν, we say φ is *valid* on the topological space $\langle W, \tau \rangle$ (written $\langle W, \tau \rangle \models \varphi$).

2.2   Practical Workings

From the perspective of modal logic, the topological semantics proposed above looks quite different from the standard Kripke-style labeled transition systems. And from the perspective of topology, it is not clear what can be expressed in the modal language $\mathcal{L}$. To sharpen intuitions on both ends, let us briefly pause to get a sense of the practical workings of the language.

At first glance, it may appear that little of mathematical interest can be expressed in $\mathcal{L}$. For example, there is no obvious way to express familiar topological properties such as connectedness, being Hausdorff, etc.. And indeed the results that follow lend support to this belief. They show, for example, that no formula of $\mathcal{L}$ can distinguish topological structures such as $\langle \mathbb{Q}, \tau \rangle$ and $\langle \mathbb{R}, \tau \rangle$, or even structures as different as $\langle \mathbb{R}, \tau \rangle$ and a finite transitive and reflexive tree equipped with the successor topology. (So it is true that connectedness or being Hausdorff can not be expressed in $\mathcal{L}$.)

On the other hand, even though many topological properties are undefinable, what formulas can be true at a point quickly becomes quite complicated. Just how complicated becomes clear below when we try to construct a model $\langle \mathbb{R}, \tau, \nu \rangle$ which contains a real number modeling an arbitrary consistent **S4** formula ψ. But for the moment let us pause to consider a few simple examples of formulas which can be made true at points on the real line.

Definition 2.6. A logic **L** is said to be *complete* with respect to a topological space $\langle W, \tau \rangle$ if $\vdash_{\mathbf{L}} \varphi \Leftrightarrow \langle W, \tau \rangle \models \varphi$.

Much of what follows will be dedicated to *proving* topological completeness results for **S4** and its extensions, but for now let us make use of a classical result.

Theorem 2.7. **S4** is complete with respect to $\langle \mathbb{R}, \tau \rangle$.

Definition 2.6 and **Theorem 2.7** taken together tell us that if ψ is an **S4**-consistent formula, then ψ can be made true at some point on the real line.

First consider an easy example, the **S4**-consistent formula $\Diamond p \wedge \neg p$. By completeness, there exists a valuation ν such that $\langle \mathbb{R}, \tau, \nu \rangle, x \models \Diamond p \wedge \neg p$ for

some $x \in \mathbb{R}$. And indeed it is not hard to think of such a v and x. For instance, if $v(p) = \{1/x \mid x > 0\}$, then $\langle \mathbb{R}, \tau, v \rangle, 0 \models \Diamond p \wedge \neg p$.

A more interesting example is the **S4**-consistent formula $\Box(\Diamond p \wedge \Diamond \neg p)$. This formula says that every point in a given open neighborhood has both p and $\neg p$ sequences approaching it. Although at first glance it might seem difficult to describe a valuation which makes this formula true at a point in $\langle \mathbb{R}, \tau \rangle$, there is actually a quite natural one. The valuation $v(p) = \mathbb{Q}$ makes $\Box(\Diamond p \wedge \Diamond \neg p)$ true at every point in $\langle \mathbb{R}, \tau \rangle$.

Both of the examples considered above involve only a single **S4**-consistent formula containing one propositional letter. However, by completeness it is guaranteed that every finite consistent collection of **S4** formulas containing any number of propositional letters can be made true together at a single point of $\langle \mathbb{R}, \tau \rangle$. This is somewhat more difficult to imagine. Now that the inner workings of the language have been explored a bit, let us see what properties can be established about the language.

2.3 The Historical Results

In "The Algebra of Topology," McKinsey and Tarski [10] examine the relationship between closure algebras and topological spaces. A closure algebra is a five-tuple $\langle K, \cup, \cap, -, C \rangle$ where

- K is a boolean algebra with respect to $\cup, \cap, -$
- If $x \in K$, then Cx is in K
- If $x \in K$, then $x \subseteq Cx$
- If $x \in K$, then $CCx = Cx$
- If x and y are in K, then $C(x \cup y) = Cx \cup Cy$
- $C\emptyset = \emptyset$

Clearly every topological space $\langle X, \tau \rangle$ can be viewed as a closure algebra $C(X) = \langle \mathcal{P}(X), \cup, \cap, -, C \rangle$, where C is the closure operator on $\mathcal{P}(X)$. Conversely, every closure algebra of the form $C(X)$ can be viewed as a topological space $\langle X, \tau \rangle$ such that $\langle X, \tau \rangle \models$

(i) $p \rightarrow \Diamond p$
(ii) $\Diamond \Diamond p \leftrightarrow \Diamond p$
(iii) $\Diamond(p \vee q) \leftrightarrow \Diamond p \vee \Diamond q$
(iv) $\Diamond \perp \leftrightarrow \perp$

It is a quick exercise to check $(i) - (iv)$ are equivalent to the axioms of the modal logic **S4**. It then follows immediately that if $\vdash_{\mathbf{S4}} \varphi$, then $\langle X, \tau \rangle \models \varphi$. McKinsey and Tarski's pioneering results in "The Algebra of Topology" show that for a wide range of topological spaces, the converse is also true.

Theorem 2.8. (McKinsey and Tarski) For every finite topological space $\langle X, \tau \rangle$ and every normal, dense-in-itself topological space $\langle Y, \tau \rangle$ with a countable base, there exists an embedding $C(X) \rightarrowtail C(Y)$.

From a logical perspective, the above theorem states that if there is a finite topological space $\langle X, \tau \rangle$ such that $\langle X, \tau \rangle \not\models \varphi$, then for any normal, dense-in-itself topological space $\langle Y, \tau \rangle$ with a countable base, $\langle Y, \tau \rangle \not\models \varphi$. It is known (and proven below in **Corollary 3.6**) that if $\nvdash_{\mathbf{S4}} \varphi$, then there is some finite topological space $\langle X, \tau \rangle$ such that $\langle X, \tau \rangle \not\models \varphi$. So **Theorem 2.8** establishes that **S4** is complete with respect to every normal, dense-in-itself topological space with a countable base. This is a very general result. It shows that, for example, **S4** is the complete with respect to the rationals, the reals, the Cantor Space, and the Baire Space.

2.4   Modern Extensions

From the perspective of modal logics of space, McKinsey and Tarski's original results are somewhat daunting in means and scope. In means, their arguments are by and large mathematical rather than metamathematical. In scope, their results are sufficiently general to settle almost all questions about the basic modal language $\mathcal{L}$ equipped with a topological semantics.

Thus recent investigations into modal logics of space have taken one of two directions. The first is to make the original results of McKinsey and Tarski more accessible by using the tools built up in modal logic over the past 70 years. This project seems to have started with Mints [11], where a new completeness proof of **S4** with respect to the Cantor Space is given. Mints's paper has in turn inspired several new proofs of completeness of **S4** with respect to the real line, including Mints, Zhang [12], Aiello, van Benthem, Bezhanishvili [3] and Bezhanishvili, Gehrke [4].

The other direction is to move to richer modal languages. In the past ten years, many spatial extensions of the basic modal language $\mathcal{L}$ have been put forward. The first and perhaps most natural extension of $\mathcal{L}$ to be considered was the language $\mathcal{L}^{\square FP}$, originally found in Shehtman [14]. $\mathcal{L}^{\square FP}$ contains the $\square$ operator, together with two additional unary operators F and P. In the intended semantics for $\mathcal{L}^{\square FP}$, $\square$ is interpreted as interior, and F and P are given their standard temporal interpretation as "future" and "past."

More formally, given a partial order $\langle W, < \rangle$, the standard order topology τ on $\langle W, < \rangle$ and a valuation v, the truth of a $\mathcal{L}^{\square FP}$ formula at a point $w \in W$ is defined as follows:

- $w \models F\psi$ iff $\exists w' > w$ such that $w' \models \psi$
- $w \models G\psi$ iff $\forall w' > w, w' \models \psi$

- $w \models P\psi$ iff $\exists w' < w$ such that $w' \models \psi$
- $w \models H\psi$ iff $\forall w' < w, w' \models \psi$
- $w \models \Box\psi$ iff $\exists U \in \tau$ such that $w \in U$ and $\forall w' \in U, w' \models \psi$
- $w \models \Diamond\psi$ iff $\forall U \in \tau, w \in U$ implies that $\exists w' \in U$ such that $w' \models \psi$

The language $\mathcal{L}^{\Box FP}$ has the appealing feature of combining two basic properties of mathematical objects: topological and metric structure. Plus it is topologically more expressive than $\mathcal{L}$. Since $\mathcal{L}^{\Box FP}$ will occur throughout this paper, let us pause for a moment to consider an example.

As mentioned before, it follows from the completeness of **S4** with respect to both $\langle \mathbb{R}, \tau \rangle$ and $\langle \mathbb{Q}, \tau \rangle$ that connectedness is not expressible in $\mathcal{L}$. For assume it is, and that φ expresses it. Then $\langle \mathbb{R}, \tau \rangle \models \varphi$ and $\langle \mathbb{Q}, \tau \rangle \not\models \varphi$, but by **Theorem 2.8**,

$$\langle \mathbb{R}, \tau \rangle \models \varphi \Leftrightarrow \vdash_{\mathbf{S4}} \varphi \Leftrightarrow \langle \mathbb{Q}, \tau \rangle \models \varphi.$$

So connectedness can not be expressed in **S4**. However, it is expressible in $\mathcal{L}^{\Box FP}$.

Lemma 2.9. $\langle W, \tau \rangle \models \Box\varphi \wedge F\neg\varphi \rightarrow F(\Diamond\varphi \wedge \Diamond\neg\varphi) \Leftrightarrow \langle W, \tau \rangle$ is connected.

Proof. $[\Rightarrow]$ Assume $\langle W, \tau \rangle$ is not connected. Then W can be written as the union of two non-empty disjoint open sets U and V. Without loss of generality, assume that there is some $x \in U$ and $y \in V$ such that $x < y$. Now consider the valuation $v(p) = U$.

It is clear that $\langle W, \tau, v \rangle, x \models \Box p \wedge F\neg p$. We want to show $\langle W, \tau, v \rangle, x \not\models F(\Diamond p \wedge \Diamond\neg p)$, so assume z is an arbitrary point such that $x < z$. Either $z \in U$ or $z \in V$. If $z \in U$, then $\langle W, \tau, v \rangle, z \models \Box p$ and if $z \in V$, then $\langle W, \tau, v \rangle, z \models \Box\neg p$. It follows that $\langle W, \tau, v \rangle, x \models G(\Box p \vee \Box\neg p)$.

$[\Leftarrow]$ Assume $\langle W, \tau \rangle \not\models \Box\varphi \wedge F\neg\varphi \rightarrow F(\Diamond\varphi \wedge \Diamond\neg\varphi)$. Then there exists a valuation v and point $x \in W$ such that $\langle W, \tau, v \rangle, x \models \Box\varphi \wedge F\neg\varphi$ and $\langle W, \tau, v \rangle, x \models G(\Box\varphi \vee \Box\neg\varphi)$. Let

$X = \{y \mid y < x\}$
$Y = \{z \mid z > x\}$
$Z = v(\Box\varphi)$
$V = v(\Box\neg\varphi)$

Then X and Y are open sets and Z and V are the unions of open sets. It follows that $U = X \cup Z$ and $T = V \cap Y$ are two disjoint open sets whose union is W. So $\langle W, \tau \rangle$ is not connected. □

As a note before moving on, one surprising feature of moving to richer modal languages is how difficult the problems become, even when seemingly increasing the expressive power of the language only slightly. For example, $\mathcal{L}^{\Box FP}$ does not seem like a particularly strong extension of $\mathcal{L}$. Furthermore, it follows from Rabin's Theorem that $\langle \mathbb{R}, \tau \rangle$ is axiomatizable in $\mathcal{L}^{\Box FP}$. Yet no one has been able to establish a complete axiomatization. Even the $\mathcal{L}^{\Box FP}$ axiomatization of $\langle \mathbb{Q}, \tau \rangle$ in Shehtman [14] is quite difficult, and something of an encyclopedia of modal logic. It seems that by adding only a small amount more expressive power, the proofs swiftly move beyond strictly logical insight.

2.5   Modal Techniques

By and large, the techniques used in investigating modal logics of space have been model-theoretic in nature. To establish completeness of a logic **L** with respect to a topological space $\langle X, \tau \rangle$, the most common strategy has been to find a convenient class of Kripke frames for which the logic is complete, and establish a modal equivalence between an arbitrary member of this class and the desired topological space.

For example, Aiello, van Benthem, Bezhanishvili [3] uses the fact that **S4** is complete with respect to the class of all finite transitive and reflexive trees to show **S4** is complete with respect to the Cantor Space. In outline, the proof the proceeds as follows:

- If $\nvdash_{\mathbf{S4}} \varphi$, by completeness φ can be refuted on some finite transitive and reflexive tree $\langle W, R \rangle$.
- A labeling is devised between branches of the Cantor space and points in $\langle W, R \rangle$ such that every node of $\langle W, R \rangle$ is modally equivalent to some branch of the Cantor Space.
- So the point on $\langle W, R \rangle$ which refutes φ is modally equivalent to some branch of the Cantor Space.
- So φ in turn can be refuted on the Cantor Space.

In the last chapter we will discuss this example in greater depth.

The line of reasoning sketched above has been used in Shehtman [14] , Mints [11], Aiello, van Benthem, Bezhanishvili [3], and van Benthem, Bezhanishvili, ten Cate, Sarenac, among other places, to give streamlined topological completeness proofs. However, the approach is not without difficulty. For one thing, when working in richer modal languages, where more topological properties can be expressed, the Kripke frame characterizations of the logic can become more difficult to establish.

On the other hand, even in cases like **S4** where the logic has many simple Kripke-frame characterizations, it still may be difficult to transfer topological structure to these Kripke frames. For example, no one has ever found a simple way to transfer the topological structure of $\langle \mathbb{R}, \tau \rangle$ to an **S4** Kripke frame. Indeed, the proofs of this fact in Aiello, van Benthem, Bezhanishvili [3] and Mints [11] work even more indirectly, first establishing a modal equivalence of the real line to some subset of the Cantor Space, then establishing a modal equivalence between this subset of the Cantor Space and an arbitrary **S4** Kripke frame. The procedure of sending points on the real line to branches of the Cantor Space to points in a Kripke model, all the while preserving modal equivalence, ends up requiring some rather fancy combinatorics.

The above difficulties raise the question of whether working indirectly through Kripke frames is always the best way to proceed. Instead of first trying to find a Kripke frame characterization of the logic and then transferring topological structure to these Kripke frames, perhaps it is better to try and think of modal techniques which work directly on the desired topological space in order to establish completeness.

2.6   The Construction Method

Curiously, there is an existing modal technique which operates directly on mathematical structures rather than indirectly through Kripke frames, but it has yet to be applied to topological completeness proofs. This technique was developed by de Jongh and Veltman [7] and Burgess [5], among others, in the 1980s to overcome precisely the difficulties described above. In its original incarnation, the technique was used for axiomatizing structures such as $\langle \mathbb{N}, < \rangle$, $\langle \mathbb{Q}, < \rangle$ and $\langle \mathbb{R}, < \rangle$ in the basic temporal language $\mathcal{L}^{FP}$.

As with their modern-day topological counterparts, the original completeness proofs for $\langle \mathbb{N}, < \rangle$, $\langle \mathbb{Q}, < \rangle$ and $\langle \mathbb{R}, < \rangle$ in the temporal language $\mathcal{L}^{FP}$ (see Segerberg [13]) provided a characterization of the relevant logic in terms of Kripke frames, and then established a modal equivalence between the desired mathematical structure and its class of Kripke frames. And again as above, the expressive power of the temporal language resulted in Kripke frame characterizations which were not very nice and, as a result, somewhat technical proofs.

Faced with these difficulties, de Jongh and Veltman (among others) developed a more direct syntactic approach towards establishing completeness. The basic idea was to construct the desired mathematical structure in stages, at each stage associating every point added in the construction with a set of formulas which will be true at that point in the model when the construction is finished. If the construction procedure is carried out correctly, there will be a harmony between the set of formulas associated with a point and the formulas which are actually true at that point when the construction is finished. Provided the sets of formulas to start the construction are selected judiciously, the structure will then serve as a model of the desired state of affairs. Let us illustrate this procedure with our first completeness proof, following an argument of de Jongh and Veltman [7].

2.7 $\mathbf{Q_{FP}}$ on $\langle \mathbb{Q}, < \rangle$

Consider the logic $\mathbf{Q_{FP}}$ containing the axioms:

$G(\psi \rightarrow \varphi) \rightarrow (G\psi \rightarrow G\varphi)$
$H(\psi \rightarrow \varphi) \rightarrow (H\psi \rightarrow H\varphi)$
$\varphi \rightarrow GP\varphi$
$\varphi \rightarrow HF\varphi$
$GG\varphi \rightarrow G\varphi$
$G\varphi \rightarrow GG\varphi$
$F\varphi \rightarrow G(\varphi \vee P\varphi \vee F\varphi)$
$P\varphi \rightarrow H(\varphi \vee P\varphi \vee F\varphi)$
$F\top$
$P\top$

and proof rules:

$$\frac{\varphi, \varphi \rightarrow \psi}{\psi} \qquad \frac{\varphi}{G\varphi} \qquad \frac{\varphi}{H\varphi}$$

Theorem 2.10. $\mathbf{Q_{FP}}$ is complete with respect to $\langle \mathbb{Q}, < \rangle$.

> *Proof.* To establish the direction $\vdash_{\mathbf{Q_{FP}}} \varphi \Rightarrow \langle \mathbb{Q}, < \rangle \models \varphi$ it is necessary to show that all of the axioms of $\mathbf{Q_{FP}}$ are valid on $\langle \mathbb{Q}, < \rangle$, and that the proof rules preserve validity on $\langle \mathbb{Q}, < \rangle$. This involves some routine fact checking, and will be left to the reader. This direction of completeness will be omitted in all further proofs. To show the converse direction, assume $\nvdash_{\mathbf{Q_{FP}}} \varphi$. We will construct $\langle \mathbb{Q}, < \rangle$ (modulo isomorphism) in such a way that $\langle \mathbb{Q}, < \rangle \nvDash \varphi$.

Definition 2.11. A set of formulas Γ is called *maximally consistent* if Γ is consistent and no set of formulas properly containing Γ is consistent.

Fact 2.12. *Every consistent set of formulas Γ can be extended to a maximally consistent set of formulas Γ'. Furthermore, Γ' is maximally consistent if and only if for any formula φ, either $\varphi \in \Gamma'$ or $\neg\varphi \in \Gamma'$ but not both.*

Definition 2.13. Let Γ_q and $\Gamma_{q'}$ be $\mathbf{Q_{FP}}$ maximally consistent sets. We say $\Gamma_q < \Gamma_{q'}$ if $G\varphi \in \Gamma_q \Rightarrow \varphi \in \Gamma_{q'}$

Fact 2.14. $\Gamma_q < \Gamma_{q'} < \Gamma_{q''} \Rightarrow \Gamma_q < \Gamma_{q''}$.

Fact 2.15. $\Gamma_q < \Gamma_{q'}$ *and* $\Gamma_q < \Gamma_{q''} \Rightarrow \Gamma_{q'} < \Gamma_{q''}$ *or* $\Gamma_{q'} = \Gamma_{q''}$ *or* $\Gamma_{q''} = \Gamma_{q'}$.

Fact 2.16.   *The following are equivalent:*

(1) $\Gamma_q < \Gamma_{q'}$
(2) $\varphi \in \Gamma_{q'} \Rightarrow F\varphi \in \Gamma_q$
(3) $H\varphi \in \Gamma_{q'} \Rightarrow \varphi \in \Gamma_q$
(4) $\varphi \in \Gamma_{q'} \Rightarrow P\varphi \in \Gamma_q$

The construction of $\langle \mathbb{Q}, < \rangle$ will proceed in stages. Each stage n will consist of:

(i)   a finite set $Q_n = \{q_0, q_1, \dots, q_k\}$
(ii)   an assignment of an $\mathbf{Q_{FP}}$ maximally consistent set Γ_{q_i} to each $q_i \in Q_n$
(iii)   a linear ordering $<$ on Q_n such that $q_j < q_h \Rightarrow \Gamma_{q_j} < \Gamma_{q_h}$

Let $\varphi_0 \varphi_1 \varphi_2 \dots$ be an enumeration of all F and P formulas in which each formula is repeated infinitely many times. The construction will proceed as follows:

- **Stage 0:** Let $Q_0 = \{q_0\}$ and associate q_0 with Γ_{q_0}, a $\mathbf{Q_{FP}}$ maximally consistent extension of $\{\neg\varphi\}$.

- **Stage 3n+1:** For all $q, q'' \in Q_{3n}$ such that q'' is the immediate successor of q in Q_{3n}, add a new point q' in between q and q'' and associate q' with $\Gamma_{q'}$, a $\mathbf{Q_{FP}}$ maximally consistent extension of $\Gamma = \{\psi \mid G\psi \in \Gamma_q\} \cup \{F\varphi \mid \varphi \in \Gamma_{q''}\}$. Let Q_{3n+1} equal Q_{3n} plus the points added by this procedure, and let $<_{3n+1}$ be $<_{3n}$ extended in the obvious way to include these additional points.

- **Stage 3n+2:** Let $F\varphi$ be the next F formula in the enumeration and let q be the greatest element in the Q_{3n+1} ordering such that $F\varphi \in \Gamma_q$ and for all $q^* \in Q_{3n+1}$ such that $q < q^*$, $\neg\varphi \in \Gamma_{q^*}$. If such a q exists, add a new point q' immediately after q and associate q' with $\Gamma_{q'}$, a $\mathbf{Q_{FP}}$ maximally consistent extension of $\Gamma' = \{\varphi\} \cup \{\psi \mid G\psi \in \Gamma_q\}$. Let Q_{3n+1} be equal to Q_{3n} plus any points added by this procedure, and let $<_{3n+1}$ be $<_{3n}$ extended in the obvious way.

- **Stage 3n+ 3.** Symmetric to $3n + 2$ taking the first P formula in the enumeration that hasn't been used.

In order to ensure that our eventual model $\langle Q, <, \nu \rangle$ turns out correctly, it must be checked that Conditions $(i) - (iii)$ above are met at every stage in the construction. As an illustrative example, we will check that condition (*iii*) is met at **Stage 3n+2**. **Fact 2.14** states that $<$ is transitive, so we only need to check that $\Gamma_q < \Gamma_{q'} < \Gamma_{q''}$, where q'' is the immediate successor of q in the $<_{3n+1}$ ordering, and q' is a new point added between q and q'' at **Stage 3n+2**.

It follows from the definition of $\Gamma_{q'}$ that $\Gamma_q < \Gamma_{q'}$. However, if q has an immediate successor q'' in Q_{3n+1}, the definition of $\Gamma_{q'}$ does not immediately guarantee that $\Gamma_{q'} < \Gamma_{q''}$. On the other hand, it does follow from **Fact 2.15** that $\Gamma_{q'} < \Gamma_{q''}$ or $\Gamma_{q''} = \Gamma_{q'}$ or $\Gamma_{q''} < \Gamma_{q'}$. If $\Gamma_{q''} = \Gamma_{q'}$ holds, then $\varphi \in \Gamma_{q''}$ and this contradicts the assumption that $\neg\varphi \in \Gamma_{q^*}$ for all $q <_{3n+1} q^*$. Furthermore, if $\Gamma_{q''} < \Gamma_{q'}$, then $F\varphi \in \Gamma_{q''}$ and this contradicts the assumption that q is the greatest element in the $<_{3n+1}$-ordering such that $F\varphi \in \Gamma_q$. So it must be the case that $\Gamma_{q'} < \Gamma_{q''}$, as required.

To finish off the construction, let $\langle Q, < \rangle = \bigcup\{\langle Q_n, <_n \rangle \mid n \in \mathbb{N}\}$. We first want to show our constructed frame $\langle Q, < \rangle$ can be used to refute φ. As suggested before, the fundamental idea behind the construction method is that it is possible to build mathematical structures in such a way that the formulas associated with a point in the construction are equivalent to the set of formulas that are true at that point when the construction is finished. However, as of yet no connection has been made between the syntactic objects associated with each point and the semantic notion of truth at a point. It is the valuation v which enables us to make this connection. To get the ball rolling, we will choose v such that the truth of a propositional letter at a point q corresponds to its membership in Γ_q. In other words, let $v(p) = \{q \mid p \in \Gamma_q\}$. Now let us check that we have performed the construction in a such a way that this harmony extends to all formulas φ.

Lemma 2.17. $\langle Q, <, v \rangle, q \models \varphi \Leftrightarrow \varphi \in \Gamma_q$.

Proof.   By induction on the complexity of φ.

- $\varphi = p$.
  By the definition of v, $q \models p \Leftrightarrow p \in \Gamma_q$.

- $\varphi = \psi \wedge \chi$.

  $q \models \psi \wedge \chi \Leftrightarrow q \models \psi$ and $q \models \chi$
  $q \models \psi$ and $q \models \chi \Leftrightarrow$ (by induction hypothesis) $\psi \in \Gamma_q$ and $\chi \in \Gamma_q$
  $\psi \in \Gamma_q$ and $\chi \in \Gamma_q \Leftrightarrow$ (by maximal consistency) $\psi \wedge \chi \in \Gamma_q$.

- $\varphi = \neg\psi$

  $q \models \neg\psi \Leftrightarrow q \not\models \psi$
  $q \not\models \psi \Leftrightarrow$ (by induction hypothesis) $\psi \notin \Gamma_q$
  $\psi \notin \Gamma_q \Leftrightarrow$ (by maximal consistency) $\neg\psi \in \Gamma_q$.

- $\varphi = F\psi$

  [$\Leftarrow$] Assume $F\psi \in \Gamma_q$. Then by construction there is a stage j such that $q \in Q_j$ and $F\psi$ is being treated. At stage $j + 1$, there is a $q' \in Q_{j+1}$ such that $q < q'$ and $\psi \in \Gamma_{q'}$. It then follows from the inductive hypothesis that $q' \models \psi$, so $\langle Q, <, v\rangle, q \models F\psi$.

  [$\Rightarrow$] Assume $\langle Q, <, v\rangle, q \models F\psi$. Then there is some $q < q'$ such that $\langle Q, <, v\rangle, q' \models \psi$. So it follows from the inductive hypothesis that $\psi \in \Gamma_q$. Since $q < q'$, we know by Condition (*iii*) on stages that $\Gamma_q \prec \Gamma_{q'}$. So $F\psi \in \Gamma_q$.

- $\varphi = G\psi$.

  [$\Leftarrow$] In this case it is easier to argue contrapostively. So assume $G\psi \notin \Gamma_q$. Then by maximal consistency $F\neg\psi \in \Gamma_q$ and it follows from "[$\Leftarrow$]" above that $\langle Q, <, v\rangle, q \models F\neg\psi$. So $\langle Q, <, v\rangle, q \not\models G\psi$.

  [$\Rightarrow$] Once again it is easier to argue contrapositively. Assume $\langle Q, <, v\rangle, q \not\models G\psi$. Then $\langle Q, <, v\rangle, q \models F\neg\psi$ and it follows from "[$\Rightarrow$]" above that $F\neg\psi \in \Gamma_q$. So by maximal consistency, $G\psi \notin \Gamma_q$.

We will leave the H and P cases to the reader, as they are quite similar to G and F. □

Now with the appropriate hindsight, it is easy to see that $\langle Q, <, v\rangle$ refutes φ. At **Stage 0** we made sure to put $\neg\varphi$ in Γ_{q_0}. So by the above lemma it follows that $\langle Q, <, v\rangle, q_0 \models \neg\varphi$.

To finish the proof, it only remains to be shown that our constructed model is isomorphic to $\langle \mathbb{Q}, <\rangle$. To do this we will make use of Cantor's Theorem.

Theorem 2.18. (Cantor) Every countable dense linear ordering without end points is isomorphic to $\mathbb{Q}$.

Lemma 2.19. $\langle Q, <\rangle$ is a countable, dense, linear order without end points.

Proof. Linearity follows immediately from Condition (*iii*) on stages, and countability from the fact that only finitely many points are added at each stage. To see $\langle Q, < \rangle$ is dense assume that $q, q'' \in Q$ such that $q < q''$. Then there is some stage Q_k such that $q, q'' \in Q_k$ and by stage Q_{k+3} a new point q' is added such that $q < q' < q''$. For right unboundedness, assume q is the greatest element in the $<_j$ ordering for some j. Then there is a stage $\ell > k$ such that the formula FT is being treated. FT is an axiom of $\mathbf{Q_{FP}}$ and so is in every $\mathbf{Q_{FP}}$ maximally consistent set. If q does not have a successor at stage ℓ there is no $q^* \in Q_\ell, q < q^*$ such that $\top \in \Gamma_{q^*}$. A new point will then be added after q at this stage. Left unboundedness is similar. $\square$

It follows from Cantor's Theorem and the above lemma that there is an isomorphism f from $\langle Q, < \rangle$ to $\langle \mathbb{Q}, < \rangle$. Assign $\langle \mathbb{Q}, < \rangle$ the valuation $v'(p) = f(v(p))$. It is easy to check that

$$\langle Q, <, v \rangle, q \models \psi \Leftrightarrow \langle \mathbb{Q}, <, v' \rangle, f(q) \models \psi$$

So $\langle \mathbb{Q}, <, v' \rangle, f(q^{\#}) \models \neg \varphi$ and $\mathbf{Q_{FP}}$ is complete with respect to $\langle \mathbb{Q}, < \rangle$. $\square$

In de Jongh, Veltman [7], the construction method presented above is used to give simplified completeness proofs for the structures $\langle \mathbb{N}, < \rangle$, $\langle \mathbb{Q}, < \rangle$ and $\langle \mathbb{R}, < \rangle$ in $\mathcal{L}^{FP}$. The simplicity of $\langle \mathbb{R}, < \rangle$ is particularly notable (and tempting) for as in the case of $\langle \mathbb{R}, \tau \rangle$, the temporal completeness proofs of $\langle \mathbb{R}, < \rangle$ working indirectly through Kripke frames are quite technical.

In the next chapter we will examine how this construction method can be applied to topological completeness proofs. We will first use the method to axiomatize $\langle \mathbb{Q}, \tau \rangle$ in the the basic modal language $\mathcal{L}$ and the more difficult combined language $\mathcal{L}^{\square FP}$. Then we will stop to consider possible extensions to $\langle \mathbb{R}, \tau \rangle$. Finally, we will examine the slightly more expressive temporal Since/ Until language on $\langle \mathbb{Q}, < \rangle$.

3   Topological Completeness Proofs

3.1   Henkin Models

Before moving on to the topological completeness proofs, let us briefly pause to correctly frame the construction method, and prove a theorem used in Section 2.4.

The construction method introduced above is best understood as a means of adding structure to Henkin models.

Definition 3.1. Consider the basic modal language $\mathcal{L}$. A *Henkin model* is a triple $M = \langle W^{\mathbf{L}}, R^{\mathbf{L}}, v^{\mathbf{L}} \rangle$ where

- $\mathbf{L}$ is a logic
- $W^{\mathbf{L}}$ is the set of all $\mathbf{L}$ maximally consistent sets
- $R^{\mathbf{L}} \subset W^{\mathbf{L}} \times W^{\mathbf{L}}$ is a binary relation such that $R^{\mathbf{L}}\Gamma_x\Gamma_y$ iff $\square\varphi \in \Gamma_x \Rightarrow \varphi \in \Gamma_y$
- $v^{\mathbf{L}}(p) = \{\Gamma_x \mid p \in \Gamma_x\}$

Both Henkin models and the construction method contain the same ingredients: maximally consistent sets, a relation determined by the syntactic structure of these sets, and a valuation which connects syntax to semantics. However, while Henkin models are rather unwieldy entities, constructed models are as well behaved as we can imagine them to be.

Yet despite their coarseness, Henkin models can be used to establish logically interesting topological completeness proofs. The following result is well known.

Theorem 3.2. **S4** is complete with respect to the class of all topological spaces.

> **Proof.** The proof will make use of the Henkin model for **S4**. Let us first define a topology on $\langle W^{\mathbf{S4}}, R^{\mathbf{S4}}, v^{\mathbf{S4}} \rangle$ as follows: let $\Delta_w = \{\Gamma_z \mid R^{\mathbf{S4}}\Gamma_w\Gamma_z\}$ and $\mathcal{B} = \{\Delta_w \mid w \in W^{\mathbf{S4}}\}$ serve as a basis for $\tau^{\mathbf{S4}}$.

Lemma 3.3. $\langle W^{\mathbf{S4}}, \tau^{\mathbf{S4}} \rangle$ is a topological space.

> **Proof.** It suffices to check the following two properties:
>
> - For any $\Delta_u, \Delta_v \in \mathcal{B}$ and any $\Gamma_x \in \Delta_u \cap \Delta_v$, there is some $\Delta_y \in \mathcal{B}$ such that $\Gamma_x \in \Delta_y \subseteq \Delta_u \cap \Delta_v$
> - For any $\Gamma_x \in W^{\mathbf{S4}}$, there is some $\Delta_y \in \mathcal{B}$ such that $\Gamma_x \in \Delta_y$

Fact 3.4. $\langle W^{\mathbf{S4}}, R^{\mathbf{S4}} \rangle$ *is transitive and reflexive.*

> Since $\langle W^{\mathbf{S4}}, R^{\mathbf{S4}} \rangle$ is reflexive the second property is immediate; for any $\Gamma_x \in W^{\mathbf{S4}}$, $\Gamma_x \in \Delta_x$. The first property is also not hard to establish. If $\Gamma_x \in \Delta_u \cap \Delta_v$ then $\Delta_x = \{\Gamma_z \mid R^{\mathbf{S4}}\Gamma_x\Gamma_z\} \subseteq \Delta_u \cap \Delta_v$. $\square$

Lemma 3.5. $\langle W^{\mathbf{S4}}, \tau^{\mathbf{S4}}, v^{\mathbf{S4}} \rangle, \Gamma_x \models \varphi \Leftrightarrow \varphi \in \Gamma_x$.

> **Proof.** By induction on the complexity of φ. The propositional and boolean cases are the same as before, and $\Diamond$ is dual to $\square$, so we will only treat the case $\varphi = \square\psi$.

[⇐] Assume $\Box\psi \in \Gamma_x$. We need to show that there is some $U \in \tau^{\mathbf{S4}}$ such that $\Gamma_x \in U$, and for every $\Gamma_y \in U$, $\Gamma_y \models \psi$. Let $U = \Delta_x = \{\Gamma_z \mid R^{\mathbf{S4}}\Gamma_x\Gamma_z\}$. Then by the definition of $R^{\mathbf{S4}}$, $\Box\psi \in \Gamma_x \Rightarrow \psi \in \Gamma_y$ for all $\Gamma_y \in U$. Furthermore, since $\langle W^{\mathbf{S4}}, R^{\mathbf{S4}}\rangle$ is reflexive, $\Gamma_x \in U$. It then follows from the inductive hypothesis that $y \models \psi$ for all $\Gamma_y \in U$. So $\Gamma_x \models \Box\psi$.

[⇒] We will argue contrapositively. Assume $\Box\psi \notin \Gamma_x$. Then by maximal consistency $\Diamond\neg\psi \in \Gamma_x$. We want to show that for every $U \in \tau^{\mathbf{S4}}$, if $\Gamma_x \in U$ then there is some $\Gamma_y \in U$ such that $\Gamma_y \models \neg\psi$. Since every open set is closed under $R^{\mathbf{S4}}$ successors, it suffices to find a point Γ_z such that $R^{\mathbf{S4}}\Gamma_x\Gamma_z$ and $\Gamma_z \models \neg\psi$. It is clear what any such Γ_z must look like: $\Gamma = \{\varphi \mid \Box\varphi \in \Gamma_x\} \cup \{\neg\psi\} \subseteq \Gamma_z$. So let Γ_z be any maximally consistent extension of Γ (we will refrain from checking that Γ is consistent, since several variants of the proof will be carried out later). It then follows that for all U such that $\Gamma_x \in U$, $\Gamma_z \in U$. So by the inductive hypothesis $\Gamma_z \models \neg\psi$. Thus $\langle W^{\mathbf{S4}}, \tau^{\mathbf{S4}}, v^{\mathbf{S4}}\rangle, \Gamma_x \not\models \Box\psi$. □

It is now easy to establish the desired completeness result. Assume $\nvdash_{\mathbf{S4}} \varphi$. We need to show there exists a topological model $\langle W, \tau, v\rangle$ such that $\langle W, \tau, v\rangle \not\models \varphi$. Since $\nvdash_{\mathbf{S4}} \varphi$, it follows by definition that $\neg\varphi$ is consistent, and so is a member of some $\mathbf{S4}$ maximally consistent set Γ_y. Then by the above lemma $\langle W^{\mathbf{S4}}, \tau^{\mathbf{S4}}, v^{\mathbf{S4}}\rangle, \Gamma_y \models \neg\varphi$. □

Corollary 3.6. **S4** is complete with respect to the class of all finite topological spaces.

Proof. Every step in the above proof goes through if we restrict the language $\mathcal{L}$ to subformulas of $\neg\varphi$. Then there are only finitely many maximally consistent sets, each of which is finite. So the topological model refuting φ will be finite. □

While the above theorem is logically quite useful, it is not immediately clear how to apply the proof method to establish completeness results for mathematically interesting structures. As it stands, Henkin models are far too coarse to prove results about specific topological spaces.

However, **Corollary 3.6** does show that Henkin models can be modified in certain ways to yield more structured completeness results. And indeed there exist far more subtle means for massaging Henkin models into a desired shape than just restricting the cardinality of the language. Yet for our purposes such subtleties are not at issue. For one of the fundamental properties of the all the mathematical structures under consideration is irreflexivity, and irreflexivity is not modally definable. This means drastic measures must be taken to guarantee that a Henkin-like model is, say, linearly ordered by a relation $<$. Rather than attempting to throw away points and remove all unwanted relations to arrive at a linear order, our construction method provides a way of building up a linear order directly. Let us see how the method can be used to establish mathematically interesting topological completeness results.

3.2   S4 on $\mathbb{Q}$

One of the the simplest topological spaces of mathematical interest is $\langle \mathbb{Q}, \tau \rangle$. The completeness of **S4** with respect to $\langle \mathbb{Q}, \tau \rangle$ follows from "The Algebra of Topology," and, in keeping with the program of simplifying McKinsey and Tarski's results, a new modal proof is given in van Benthem, Bezhanishvili, ten Cate, Sarenac. We will provide another proof using the construction method which is slightly stronger than existing results, which establish completeness rather than strong completeness.

Definition 3.7.   A logic **L** is said to be *strongly complete* with respect to a model $M = \langle W, R, \nu \rangle$ if for any set of formulas Δ and formula φ:

$$M \models \Delta \Rightarrow M \models \varphi \ (\Delta \models_M \varphi) \text{ iff } \varphi \text{ is provable in } \mathbf{L} \text{ from } \Delta \ (\Delta \vdash_\mathbf{L} \varphi).$$

Strong completeness implies completeness (let $\Delta = \emptyset$), but the converse is not true.

Theorem 3.8.   **S4** is strongly complete with respect to $\langle \mathbb{Q}, \tau \rangle$.

> **Proof.**   Assume $\Delta \nvdash_\mathbf{S4} \chi$. As in temporal case, we will give a construction of $\langle \mathbb{Q}, < \rangle$. We will then view $\langle \mathbb{Q}, < \rangle$ as the topological space $\langle \mathbb{Q}, \tau \rangle$ and show that the construction has been carried out such that $\Delta \nvDash_{\langle \mathbb{Q}, \tau \rangle} \varphi$.
>
> In this case, the construction will be divided into *stages* and *steps*. A *stage n* consists of:
>
> (i)   a finite set $Q_n = \{q_0, q_1, \ldots, q_k\}$
> (ii)   a linear ordering $<_n$ on Q_n
> (iii)   an assignment of an **S4** maximally consistent set Γ_{q_i} to each $q_i \in Q_n$
>
> Each stage $n > 0$ is divided into $k + 2$ steps. A *step j, $0 \le j \le k + 1$,* consists of:
>
> (i)   a finite set $Q_n^j = \{q_0, q_1, \ldots, q_l\}$
> (ii)   a linear ordering $<_n^j$ on Q_n^j
> (iii)   an assignment of an **S4** maximally consistent set Γ_{q_i} to each $q_i \in Q_n^j$

Note that unlike the temporal construction of $\langle \mathbb{Q}, < \rangle$, the relations $<_n$ and $<_n^j$ are independent of the syntactic structure of the maximally consistent sets. For someone accustomed to Henkin-style completeness proofs, this probably seems a bit off. But really it is no mystery. When the truth of $\mathcal{L}$ formulas is evaluated on $\langle \mathbb{Q}, \tau \rangle$, the ordering of $\langle \mathbb{Q}, < \rangle$ will play no role. During the construction we need to keep in mind the eventual topological structure and not the ordering $\langle \mathbb{Q}, < \rangle$.

With this word of advice, let us begin the construction. Let $\varphi_0 \varphi_1 \varphi_2 \ldots$ be an enumeration of all $\square$ and $\diamond$ formulas in which each formula is repeated infinitely many times.

Stage 0:   Let $Q_0 = \{q_0\}$ and associate q_0 with Γ_{q_0}, an **S4** maximally consistent extension of $\Delta \cup \{\neg \chi\}$.

Stage 2n+1:   Let $\square \varphi$ be the next $\square$ formula in the enumeration and $\{q_0, \ldots, q_k\}$ be the elements of Q_{2n} ordered by $<_{2n}$.

- **Step 0:** $Q_{2n} = Q_{2n}^0$ and $<_{2n} = <_{2n}^0$.
- **Step j+1:**   If $\square \varphi \in \Gamma_{q_j}$ let $Q_{2n}^{j+1} = Q_{2n}^j \cup \{q^*, q'\}$ and let $<_{2n}^{j+1}$ be $<_{2n}^j$ extended to include new points q^* and q' immediately before and after q_j. Associate q^* and q' with Γ_{q_j}. If $\square \varphi \notin \Gamma_{q_j}$, let $Q_{2n}^{j+1} = Q_{2n}^j$ and $<_{2n}^{j+1} = <_{2n}^j$.

Let $Q_{2n+1} = Q_{2n}^{k+1}$ and $<_{2n+1} = <_{2n}^{k+1}$.

Stage 2n+2:   Let $\diamond \varphi$ be the next $\diamond$ formula in the enumeration and $\{q_0, \ldots, q_k\}$ be the elements of Q_{2n+1} ordered by $<_{2n+1}$.

- **Step 0:** $Q_{2n+1} = Q_{2n+1}^0$ and $<_{2n+1} = <_{2n+1}^0$.
- **Step j+1:**   If $\diamond \varphi \in \Gamma_{q_j}$, let $Q_{2n+1}^{j+1} = Q_{2n+1}^j \cup \{q'\}$ and $<_{2n+1}^{j+1}$ be $<_{2n}^j$ extended to include q' immediately after q_j. Associate q' with $\Gamma_{q'}$, an **S4** maximally consistent extension of $\Gamma = \{\varphi\} \cup \{\square \psi \mid \square \psi \in \Gamma_{q_j}\}$. If $\diamond \varphi \notin \Gamma_{q_j}$, add no points at step Q_{2n+1}^{j+1}.

Let $Q_{2n+2} = Q_{2n+1}^{k+1}$ and $<_{2n+2} = <_{2n+1}^{k+1}$.

To finish off the construction, let $\langle Q, < \rangle = \bigcup \{\langle Q_n, <_n \rangle \mid n \in \mathbb{N}\}$. Clearly each stage and each step meets the conditions imposed above. The only thing to check is that the sets $\Gamma_{q'}$ used in **Stage 2n+2** are consistent. By **Fact 2.12**, it suffices to check that Γ is consistent.

Lemma 3.9. Γ is consistent.

Proof. Assume not. Then there are $\Box\psi_1, ..., \Box\psi_n \in \Gamma_{q_j}$ such that

$$\vdash_{\mathbf{S4}} \Box\psi_1 \wedge ... \wedge \Box\psi_n \wedge \varphi \rightarrow \bot$$
$$\vdash_{\mathbf{S4}} \Box\psi \rightarrow \neg\varphi \text{ where } \Box\psi = \Box(\psi_1 \wedge ... \wedge \psi_n)$$
$$\vdash_{\mathbf{S4}} \Box(\Box\psi \rightarrow \neg\varphi)$$
$$\vdash_{\mathbf{S4}} \Box\Box\psi \rightarrow \Box\neg\varphi$$
$$\vdash_{\mathbf{S4}} \Box\psi \rightarrow \Box\Box\psi$$
$$\vdash_{\mathbf{S4}} \Box\psi \rightarrow \Box\neg\varphi$$

Then since $\Box\psi \in \Gamma_{q_j}$, it follows that $\Box\neg\varphi \in \Gamma_{q_j}$. This contradicts the fact that $\Diamond\varphi \in \Gamma_{q_j}$ and Γ_{q_j} is consistent. $\square$

Before switching perspectives and viewing our constructed frame $\langle Q, < \rangle$ as a topological space, let us first check that it has the desired structure.

Lemma 3.10. $\langle Q, < \rangle$ is a countable dense unbounded linear order.

Proof. It follows from Condition *(ii)* on stages and steps that $\langle Q, < \rangle$ is linear. And again, since only finitely many points are added at every stage, $\langle Q, < \rangle$ is countable. To see $\langle Q, < \rangle$ is dense, consider two arbitrary points $q, q'' \in Q$ such that $q < q''$. Then there is some stage k such that $q, q'' \in Q_k$ and the formula $\Box\top$ is being treated. $\Box\top$ is a member of every **S4** maximally consistent set, so in particular $\Box\top \in \Gamma_q$. Thus by construction, at **Stage** $k+1$ there will be a new point q' such that $q < q' < q''$. The same argument shows $\langle Q, < \rangle$ is unbounded. $\square$

Now let us switch perspectives and view our ordered set $\langle Q, < \rangle$ as a topological space with the standard topology. Assign $\langle Q, \tau \rangle$ the expected Henkin valuation $v(p) = \{q \mid p \in \Gamma_q\}$.

Lemma 3.11. $\langle Q, \tau, v \rangle, q \models \varphi \Leftrightarrow \varphi \in \Gamma_q$.

Proof. By induction on the complexity of φ. We will only treat the case $\varphi = \Box\psi$. $[\Rightarrow]$ Assume $\Box\psi \notin \Gamma_q$. Then since Γ_q is an **S4** maximally consistent set, $\Diamond\neg\psi \in \Gamma_q$. We want to show that for every $U \in \tau$ such that $q \in U$, there is some $y \in U$ such that $y \models \neg\psi$. Let U' be an arbitrary open containing q and let q'' be a point in U' such that $q < q''$. That such a q'' exists is guaranteed by the density of $\langle Q, < \rangle$. Then there is some stage j in the construction such that $q, q'' \in Q_{j-1}$ and $\Diamond\neg\psi$ is being treated. When stage j is completed, there is a point $q' \in Q_j$ such that $q < q' < q''$ and $\neg\psi \in \Gamma_{q'}$. It then follows from the inductive hypothesis that $q' \models \neg\psi$. So $\langle Q, \tau, v \rangle, q \not\models \Box\psi$.

[⟸] Assume $\Box\psi \in \Gamma_q$. We want to show that there is some $U \in \tau$ such that $q \in U$ and for all $y \in U$, $y \models \psi$. To find such a U, consider the first stage j such that $q \in Q_{j-1}$ and $\Box\psi$ is being treated. At stage j, there is some step i where points q^{**} and q'' are added immediately before and after q and $\Gamma_{q^{**}} = \Gamma_q = \Gamma_{q''}$. We claim that for all Q_p^m, $m > i$, $p \geq j - 1$, if $q' \in Q_p^m$ and $q^{**} < q' < q''$, then $\Box\psi \in \Gamma_{q'}$. We will argue by induction on the steps of the construction.

Assume at step Q_s^{r+1}, $r + 1 > i$, $s \geq j - 1$, a point q' is added between q^{**} and q''. By inspecting the construction, it is clear that q' is added immediately before or after the point $q_r \in Q_s$. So $q^{**} \leq q_r \leq q''$ and it follows from the inductive hypothesis and the fact that $\Box\psi$ is in $\Gamma_{q^{**}}$ and $\Gamma_{q''}$ that $\Box\psi \in \Gamma_{q_r}$. If s is odd then $\Gamma_{q'} = \Gamma_{q_r}$ and if s is even then $\{\Box\psi \mid \Box\psi \in \Gamma_{q_r}\} \subseteq \Gamma_{q'}$. So $\Box\psi \in \Gamma_{q'}$.

Finally, $\Box\psi \rightarrow \psi$ is an axiom of **S4**, so for all $q' \in Q$ such that $q^{**} < q' < q''$, $\psi \in \Gamma_{q'}$. By the inductive hypothesis on the complexity formulas, for all $q' \in Q$ such that $q^{**} < q' < q''$, $q' \models \psi$. So if we let U be the open $\{y \mid q^{**} < y < q''\}$ we immediately get that $\langle Q, \tau, v \rangle, q \models \Box\psi$. ☐

To finish the proof, let f be an isomorphism between $\langle Q, < \rangle$ and $\langle \mathbb{Q}, < \rangle$. Assign $\langle \mathbb{Q}, \tau \rangle$ the valuation $v'(p) = f(v(p))$. It is easy to check that $\langle Q, \tau, v \rangle, q \models \varphi \Leftrightarrow \langle \mathbb{Q}, \tau, v' \rangle, f(q) \models \varphi$. Since we made sure to add $\Delta \cup \{\neg\chi\}$ to Γ_{q_0} at the first stage of the construction, it follows from the above lemma that $\langle Q, \tau, v \rangle, q_0 \models \Delta \cup \{\neg\chi\}$. So $\langle \mathbb{Q}, \tau, v' \rangle, f(q_0) \models \Delta \cup \{\neg\chi\}$ and **S4** is strongly complete with respect to $\langle \mathbb{Q}, \tau \rangle$. ☐

3.3 Remarks on the Construction Method

The essential feature of the above construction is that once a $\Box\varphi$ formula is treated at a step, a $\Box\varphi$ stretch is created which is preserved for the rest of the construction. Since this is the most stringent requirement on the construction, as might be guessed, there is considerable freedom in how formulas are satisfied at a point in $\langle \mathbb{Q}, \tau \rangle$.

To see this, consider the construction given above. If $\Diamond\varphi \in \Gamma_q$, then maximally consistent sets containing φ are added arbitrarily close to q *on the right*. So if $\langle \mathbb{Q}, \tau, v' \rangle, q \models \Diamond\varphi$, there is guaranteed to be a sequence approaching from the right such that every point models φ. However, we could have just as well selected for a sequence from the left, or sequences from both the right and the left. Actually, this last alternative has an interesting property: for all formulas φ, $v'(\Box\varphi) = X$ has no least upper bound. For assume X does have a least upper bound q^*. Then $\Box\varphi \notin \Gamma_{q^*}$, so $\Diamond\neg\varphi \in \Gamma_{q^*}$. However, by construction there is a sequence of points modeling $\neg\varphi$ approaching q^* from left, so q^* is not the least upper bound of $\nu'(\Box\varphi)$ after all.

This flexibility in how formulas are to be satisfied on the desired topological space is a special feature of the construction method. When proving topological completeness results indirectly through Kripke models, the starting point is an arbitrary **S4** model M refuting a formula φ. The end result is that every point in the desired topological space is modally equivalent to some point in M. From this fact it follows that the desired topological space refutes φ, but we don't have any idea *how* it does this.

The flexibility afforded by the construction method might lead one to be believe that it has an advantage over the model-theoretic approaches, at least in certain cases. For example, by the above result we know $\langle \mathbb{Q}, \tau \rangle$ is a counter-model to an arbitrary non-theorem φ of **S4**. However, lacking any additional insight about how φ is refuted on $\langle \mathbb{Q}, \tau, \nu \rangle$, it seems rather hopeless to extend $\langle \mathbb{Q}, \tau, \nu \rangle$ to a proposed countermodel $\langle \mathbb{R}, \tau, \nu' \rangle$ refuting φ. But perhaps by constructing $\langle \mathbb{Q}, \tau \rangle$ so that φ is refuted in a particular way, it might be possible to extend $\langle \mathbb{Q}, \tau, \nu \rangle$ to a successful countermodel on $\langle \mathbb{R}, \tau, \nu' \rangle$. Indeed, as we shall see shortly, extending a $\mathbf{Q_{FP}}$ countermodel $\langle \mathbb{Q}, <, \nu \rangle$ to an $\mathbf{R_{FP}}$ countermodel $\langle \mathbb{R}, < \nu' \rangle$ in the basic temporal language works quite naturally.

3.4 $\mathbf{Q_{\Box FP}}$ on $\mathbb{Q}$

Before considering possible extensions of $\mathbb{Q}$ to $\mathbb{R}$ in both temporal and topological languages, let us first examine the rationals in the combined temporal and topological language $\mathcal{L}^{\Box FP}$. As mentioned in the previous chapter, $\mathcal{L}^{\Box FP}$ was one of the first spatial extensions of the basic modal language to be considered. The language can be traced back to Shehtman [14], where an axiomatization of $\langle \mathbb{Q}, \tau \rangle$ is given. (For simplicity and to highlight the topological structure, we write $\langle \mathbb{Q}, \tau \rangle$ rather than $\langle \mathbb{Q}, <, \tau \rangle$ or $\langle \mathbb{Q}, < \rangle$.) The completeness proof for $\langle \mathbb{Q}, \tau \rangle$ given in Shehtman [14] works indirectly through Kripke frames, and turns out to be something of an encyclopedia of modal logic. All of the potential difficulties of applying the model-theoretic method do in fact appear, especially in providing a characterization of the logic and transferring structure to a linear frame. This is one case where using the construction method improves matters significantly and yields the slightly more robust result of strong completeness.

Let $\mathbf{Q_{\Box FP}}$ be the logic **S4** + $\mathbf{Q_{FP}}$ + :

$F\varphi \rightarrow \Box F\varphi$
$P\varphi \rightarrow \Box P\varphi$
$\Box\varphi \rightarrow F\varphi$
$\Box\varphi \rightarrow P\varphi$
$\Diamond\varphi \rightarrow \varphi \vee F\varphi \vee P\varphi$

$$\Box\varphi \land G\varphi \rightarrow \Box G\varphi$$
$$\Box\varphi \land H\varphi \rightarrow \Box H\varphi$$

Theorem 3.12. $\mathbf{Q_{\Box FP}}$ is strongly complete with respect to $\langle \mathbb{Q}, \tau \rangle$.

Proof. Assume $\Delta \nvdash_{\mathbf{Q_{\Box FP}}} \chi$. As before, we will construct a frame $\langle Q, < \rangle \cong \langle \mathbb{Q}, < \rangle$ in steps and stages such that $\langle Q, \tau \rangle$ refutes $\Delta \cup \{\varphi\}$. However, since the ordering of $\langle Q, < \rangle$ will play a role in evaluating the truth of $\mathcal{L}^{\Box FP}$ formulas in $\langle \mathbb{Q}, \tau \rangle$, this construction will require some relation between the maximally consistent sets and $<$. We will build this interaction into the definitions of a stage and a step.

A stage n consists of:

(i)   a finite set $Q_n = \{q_0, q_1, ..., q_k\}$
(ii)   a linear ordering $<_n$ on Q_n such that $q_j <_n q_l \Rightarrow \Gamma_{q_j} < \Gamma_{q_l}$
(iii)   an assignment of a $\mathbf{Q_{\Box FP}}$ maximally consistent set Γ_{q_i} to each $q_i \in Q_n$

Each stage n is divided into $\leq k + 2$ steps. A step j, $0 \leq j \leq k + 1$, consists of:

(i)   a finite set $Q_n^j = \{q_0, q_1, ..., q_l\}$
(ii)   a linear ordering $<_n^j$ on Q_n^j such that $q_h <_n^j q_i \Rightarrow \Gamma_{q_h} < \Gamma_{q_i}$
(iii)   an assignment of a $\mathbf{Q_{\Box FP}}$ maximally consistent set Γ_{q_i} to each $q_i \in Q_n^j$

Once again, the salient feature of the construction will be that once a $\Box\psi$ formula is treated at a step, that a $\Box\psi$ stretch is created which is preserved for the remainder of the construction. However, in the present case, preserving $\Box\psi$ stretches requires some attention. For example, assume the point q is in what we intend to be a $\Box\psi$ stretch and that $H\varphi$ is in Γ_q. Furthermore, assume at some stage we add a point q'' immediately after q and $P(\neg\varphi \land \Diamond\neg\psi)$ is added to $\Gamma_{q''}$. Nothing so far prevents this situation from occurring. We then know by the ordering relation $\prec$ that $H\varphi$ must be in all Γ_{q^*} such that $q^* < q$. So in order to ensure the truth of $P(\neg\varphi \land \Diamond\neg\psi)$ at q'' when the construction is finished, a point q' must be added in between q and q'' such that $\Diamond\neg\psi \in \Gamma_{q'}$. This ruins our prospective $\Box\psi$ stretch. Fortunately, guarding against this sort of occurrence alone is sufficient to preserve $\Box\psi$ stretches.

Definition 3.13. We say $\Gamma_q <_{\Box\psi} \Gamma_{q'}$ if

(1) $\Gamma_q < \Gamma_{q'}$
(2) $\Box\psi \in \Gamma_q, \Gamma_{q'}$

(3) $H\varphi \in \Gamma_q \Rightarrow H(\varphi \vee \Box\psi) \in \Gamma_{q'}$
(4) $G\varphi \in \Gamma_{q'} \Rightarrow G(\varphi \vee \Box\psi) \in \Gamma_q$

Proposition 3.14. $\Gamma_{q^*} <_{\Box\psi} \Gamma_q <_{\Box\psi} \Gamma_{q'} \Rightarrow \Gamma_{q^*} <_{\Box\psi} \Gamma_{q'}$.

Assume $\Gamma_{q^*} <_{\Box\psi} \Gamma_q <_{\Box\psi} \Gamma_{q'}$. It is easy to see Conditions (1) and (2) of $\Gamma_{q^*} <_{\Box\psi} \Gamma_{q'}$ are met. So assume $H\varphi \in \Gamma_{q^*}$. By assumption $\Gamma_{q^*} <_{\Box\psi} \Gamma_q$, so $H(\varphi \vee \Box\psi) \in \Gamma_q$. Furthermore, $\Gamma_q <_{\Box\psi} \Gamma_{q'}$, so $H((\varphi \vee \Box\psi) \vee \Box\psi) \in \Gamma_{q'}$. It then follows from basic propositional logic and maximal consistency that $H(\varphi \vee \Box\psi) \in \Gamma_{q'}$. Condition (4) is similar.   □

Similarly we have:

Fact 3.15. $\Gamma_q <_{\Box\psi} \Gamma_{q'}$ and $\Gamma_q <_{\Box\varphi} \Gamma_{q'} \Rightarrow \Gamma_q <_{\Box(\psi \wedge \varphi)} \Gamma_{q'}$.

Armed with our new condition $<_{\Box\psi}$ to preserve $\Box\psi$ stretches, we are ready to begin the construction. Let $\varphi_0\varphi_1\varphi_2...$ be an enumeration of all $F, P, \Box$ and $\Diamond$ formulas in which each formula is repeated infinitely many times.

Stage 0: Let $Q_0 = \{q_0\}$ with which we associate Γ_{q_0}, a $\mathbf{Q_{\Box FP}}$ maximally consistent extension of $\Delta \cup \{\neg\chi\}$.

Stage $4n + 1$: Let $\Box\varphi$ be the next $\Box$ formula in the enumeration and let $\{q_0, ..., q_k\}$ be the elements in Q_{4n} ordered by $<_{4n}$.

- **Step 0:** $Q_{4n} = Q_{4n}^0$ and $<_{4n}=<_{4n}^0$
- **Step j+1:**   If $\Box\varphi \in \Gamma_{q_j}$, let q^{**} be the immediate predecessor of q_j in the $<_{4n}^j$ ordering, provided one exists. Add a new point q^* immediately before q_j and associate q^* with Γ_{q^*}, a $\mathbf{Q_{\Box FP}}$ maximally consistent extension of:

$$\Gamma^* = \{\Box\varphi\} \cup \{\Box\psi' \mid \Gamma_{q^{**}} <_{\Box\psi'} \Gamma_{q_j}\} \cup$$
$$\{\alpha' \mid H\alpha' \in \Gamma_{q_j}\} \cup \{P\beta' \mid \beta' \in \Gamma_{q^{**}}\} \cup$$
$$\{P\gamma' \mid P(\gamma' \wedge \neg\Box\varphi) \in \Gamma_{q_j}\} \cup \{G\zeta' \mid G(\zeta' \vee \Box\varphi) \in \Gamma_{q_j}\}$$

Similarly, let q'' be the immediate successor of q_j in the $<_{4n}^j$ ordering, provided one exists. Add a new point q' immediately after q_j and associate q' with $\Gamma_{q'}$, a $\mathbf{Q_{\Box FP}}$ maximally consistent extension of:

$$\Gamma' = \{\Box\varphi\} \cup \{\Box\psi \mid \Gamma_{q_j} <_{\Box\psi} \Gamma_{q''}\} \cup$$
$$\{\alpha \mid G\alpha \in \Gamma_{q_j}\} \cup \{F\beta \mid \beta \in \Gamma_{q''}\} \cup$$
$$\{H(\gamma \vee \Box\varphi) \mid H\gamma \in \Gamma_{q_j}\} \cup \{F\zeta \mid F(\zeta \wedge \neg\Box\varphi) \in \Gamma_{q_j}\}$$

Let Q_{4n}^{j+1} be Q_{4n}^j plus any points added by the above procedure, and $<_{4n}^{j+1}$ be $<_{4n}^j$ extended in the obvious way to include these points.

Let $Q_{4n+1} = Q_{4n}^{k+1}$ and $<_{4n+1} = <_{4n}^{k+1}$.

The definitions of Γ^* and Γ' ensure that for all $q \in Q_{4n}$, if $\Box\varphi \in \Gamma_q$ then at Stage 4n+1 q will have an immediate predecessor q^* and immediate successor q' such that $\Gamma_{q*} <_{\Box\varphi} \Gamma_q <_{\Box\varphi} \Gamma_{q'}$. It only remains to be checked that Γ' is consistent.

Lemma 3.16. Γ' is consistent.

Proof.   Assume not. Then by **Fact 3.15** there is a formula $\Box\psi$ and formulas $G\alpha, H\gamma_1, .., H\gamma_m, F(\zeta_1 \wedge \neg\Box\varphi), .., F(\zeta_p \wedge \neg\Box\varphi) \in \Gamma_{q_j}$ and $\beta_1, .., \beta_k \in \Gamma_{q''}$ such that:

$\vdash_{\mathbf{Q_{\Box FP}}}$ $\Box\varphi \wedge \Box\psi \wedge \alpha \wedge F\beta_1 \wedge ... \wedge F\beta_k \wedge H(\gamma_1 \vee \Box\varphi) \wedge ...$
$\wedge H(\gamma_m \vee \Box\varphi) \wedge F\zeta_1 \wedge ... \wedge F\zeta_p \rightarrow \bot$

$\vdash_{\mathbf{Q_{\Box FP}}}$ $\alpha \rightarrow \neg(\Box\varphi \wedge \Box\psi \wedge F\beta_1 \wedge ... \wedge F\beta_k \wedge H(\gamma_1 \vee \Box\varphi) \wedge ...$
$\wedge H(\gamma_m \vee \Box\varphi) \wedge F\zeta_1 \wedge ... \wedge F\zeta_p)$

$(*) \vdash_{\mathbf{Q_{\Box FP}}}$ $G\alpha \rightarrow G\neg(\Box\varphi \wedge \Box\psi \wedge F\beta_1 \wedge ... \wedge F\beta_k \wedge H(\gamma_1 \vee \Box\varphi) \wedge$
$... \wedge H(\gamma_m \vee \Box\varphi) \wedge F\zeta_1 \wedge ... \wedge F\zeta_p)$

We know that:

- $\Box\varphi \rightarrow \Box\Box\varphi$ is an axiom and $\Box\varphi, \Box\psi \in \Gamma_{q_j}$, so $\Box\Box\varphi \in \Gamma_{q_j}$ and $\Box\Box\psi \in \Gamma_{q_j}$
- $\beta_1, ..., \beta_k \in \Gamma_{q''}$ and $\Gamma < \Gamma_{q''}$, so $F\beta_1, ..., F\beta_k \in \Gamma_{q''}$
- $F\beta \rightarrow \Box F\beta$ is an axiom, so $\Box F\beta_1, ..., \Box F\beta_k \in \Gamma_{q_j}$
- $\Box\varphi \in \Gamma_{q_j}$, so $\Box(\gamma_1 \vee \Box\varphi), ..., \Box(\gamma_m \vee \Box\varphi) \in \Gamma_{q_j}$
- $H\gamma_1, ..., H\gamma_m \in \Gamma_{q_j}$, so $H(\gamma_1 \vee \Box\varphi), ..., H(\gamma_m \vee \Box\varphi) \in \Gamma_{q_j}$
- $\Box\psi \wedge H\psi \rightarrow \Box H\psi$ is an axiom, so $\Box H(\gamma_1 \vee \Box\varphi), ..., \Box H(\gamma_m \vee \Box\varphi) \in \Gamma_{q_j}$
- $F\zeta_1, ..., F\zeta_p \in \Gamma_{q_j}$ and $F\zeta \rightarrow \Box F\zeta$ is an axiom, so $\Box F\zeta_1, ..., \Box F\zeta_p \in \Gamma_{q_j}$

Putting all this together it follows that:

$$\Box(\Box\varphi \wedge \Box\psi \wedge F\beta_1 \wedge ... \wedge F\beta_k \wedge H(\gamma_1 \vee \Box\varphi) \wedge ... \wedge H(\gamma_m \vee \Box\varphi) \wedge$$
$$F\zeta_1 \wedge ... \wedge F\zeta_p) \in \Gamma_{q_j}$$

and since $\Box\varphi \rightarrow F\varphi$ is an axiom:

$$F(\Box\varphi \wedge \Box\psi \wedge F\beta_1 \wedge ... \wedge F\beta_k \wedge H(\gamma_1 \vee \Box\varphi) \wedge .... \wedge H(\gamma_m \vee \Box\varphi) \wedge$$
$$F\zeta_1 \wedge ... F\zeta_p) \in \Gamma_{q_j}$$

However, $G\alpha \in \Gamma_{q_j}$, so by $(*)$:

$$G\neg(\Box\varphi \wedge \Box\psi \wedge F\beta_1 \wedge \ldots \wedge F\beta_k \wedge H(\gamma_1 \vee \Box\varphi) \wedge \ldots \wedge H(\gamma_m \vee \Box\varphi) \wedge F\zeta_1 \wedge \ldots \wedge F\zeta_p) \in \Gamma_{q_j}$$

This contradicts the consistency of Γ_{q_j}. $\Box$

Showing Γ^* is consistent is similar.

Stage $4n+2$: Let $F\varphi$ be the next F formula in the enumeration, and let $\{q_0, \ldots, q_k\}$ be the elements in Q_{4n+2} ordered by $<_{4n+2}$.

- **Step 0:** If $F\varphi \in \Gamma_{q_j}$ and $\neg\varphi \wedge G\neg\varphi \in \Gamma_{q_{j+1}}$ (or if $F\varphi \in \Gamma_{q_j}$ and $j = k$) add a new point q' immediately after q_j and associate $\Gamma_{q'}$ with a $\mathbf{Q_{\Box FP}}$ maximally consistent extension of:

$$\Lambda = \{\varphi\} \cup \{\Box\psi \mid \Gamma_{q_j} <_{\Box\psi} \Gamma_{q_{j+1}}\} \cup \{\alpha \mid G\alpha \in \Gamma_{q_j}\}$$

  Let Q^0_{4n+1} be Q_{4n+1} plus any points added by this procedure and $<^0_{4n+1}$ be $<_{4n+1}$ extended in the obvious way to include any new points.

Let $Q_{4n+2} = Q^0_{4n+1}$ and $<_{4n+2} = <^0_{4n+1}$.

Lemma 3.17. Λ is consistent.

> **Proof.** Assume not. Then there are formulas $\Box\psi$ and α in Γ_{q_j} such that:
>
> $$\vdash_{\mathbf{Q_{\Box FP}}} \varphi \wedge \Box\psi \wedge \alpha \to \bot$$
> $$\vdash_{\mathbf{Q_{\Box FP}}} \alpha \to \neg(\Box\psi \wedge \varphi)$$
> $$\vdash_{\mathbf{Q_{\Box FP}}} G\alpha \to G\neg(\Box\psi \wedge \varphi)$$
>
> Since $G\alpha \in \Gamma_{q_j}$, it follows that $G(\neg\Box\psi \vee \neg\varphi) \in \Gamma_{q_j}$. Furthermore, since $G\neg\varphi \in \Gamma_{q_{j+1}}$ and $\Gamma_{q_j} <_{\Box\psi} \Gamma_{q_{j+1}}$, it follows that $G(\Box\psi \vee \neg\varphi) \in \Gamma_{q_j}$. So:
>
> $$G(\Box\psi \to \neg\varphi) \wedge G(\neg\Box\psi \to \neg\varphi) \in \Gamma_{q_j}$$
>
> However, by basic propositional logic it then follows that $G\neg\varphi \in \Gamma_{q_j}$. This contradicts the fact that $F\varphi \in \Gamma_{q_j}$. $\Box$

The same argument given in the temporal construction of $\langle Q, < \rangle$ suffices to show that $\Gamma_{q_j} < \Gamma_{q'} < \Gamma_{q_{j+1}}$.

Stage $4n+3$: Symmetric to $4n+2$ for the next P formula in the enumeration.

Stage $4n+4$: Let $\Diamond\varphi$ be the next $\Diamond$ formula in the enumeration and let $\{q_0, \ldots, q_k\}$ be the elements in Q_{4n+3} ordered by $<_{4n+3}$.

- **Step 0:** Let $Q_{4n+3} = Q_{4n+3}^0$ and $<_{4n+3} = <_{4n+3}^0$.
- **Step $j+1$:** If $\Diamond\varphi \in \Gamma_{q_j}$ let q'' be the immediate successor and q^{**} be the immediate predecessor of q_j in the $<_{4n+3}^j$ ordering, provided these exist. Let

$$\Lambda' = \{\varphi\} \cup \{\Box\psi \mid \Gamma_{q_j} <_{\Box\psi} \Gamma_{q''}\} \cup \{\alpha \mid G\alpha \in \Gamma_{q_j}\} \cup \{F\beta \mid \beta \in \Gamma_{q''}\}$$

and

$$\Lambda^* = \{\varphi\} \cup \{\Box\psi' \mid \Gamma_{q_j} <_{\Box\psi'} \Gamma_{q^{**}}\} \cup \{\alpha' \mid H\alpha' \in \Gamma_{q_j}\} \cup$$
$$\{P\beta' \mid \beta' \in \Gamma_{q^{**}}\}$$

If $\varphi \in \Gamma_{q_j}$ add no points. If $\varphi \notin \Gamma_{q_j}$ and Λ' is consistent, then add a new point q' immediately after q_j and associate q' with a maximally consistent extension of Λ'. Otherwise, add a new point q^* immediately before q_j and associate q^* with a maximally consistent extension of Λ^*.

Let $Q_{4n+4} = Q_{4n+3}^{k+1}$ and $<_{4n+4} = Q_{4n+3}^{k+1}$.

Proposition 3.18. Either $\varphi \in \Gamma_{q_j}$, or Λ' is consistent, or Λ^* is consistent.

Proof. Assume not. Then there are formulas $\Box\psi, \Box\psi', G\alpha$, $H\alpha' \in \Gamma_{q_j}, \beta_1, ..., \beta_n \in \Gamma_{q''}$ and $\beta_1', ..., \beta_k' \in \Gamma_{q^{**}}$ such that

- $\vdash_{\mathbf{Q_{\Box FP}}} \varphi \wedge \Box\psi \wedge \alpha \wedge F\beta_1 \wedge ... \wedge F\beta_n \to \bot$
- $\vdash_{\mathbf{Q_{\Box FP}}} \varphi \wedge \Box\psi' \wedge \alpha' \wedge P\beta_1' \wedge ... \wedge P\beta_k' \to \bot$

As in the previous stages, it follows that:

- $\vdash_{\mathbf{Q_{\Box FP}}} G\alpha \to G\neg(\varphi \wedge \Box\psi \wedge F\beta_1 \wedge ... \wedge F\beta_n)$
- $\vdash_{\mathbf{Q_{\Box FP}}} H\alpha' \to H\neg(\varphi \wedge \Box\psi' \wedge P\beta_1' \wedge ... \wedge P\beta_k')$

and thus (**):

- $\varphi \notin \Gamma_{q_j}$ and
- $G\neg(\varphi \wedge \Box\psi \wedge F\beta_1 \wedge ... \wedge F\beta_n) \in \Gamma_{q_j}$ and
- $H\neg(\varphi \wedge \Box\psi' \wedge P\beta_1' \wedge ... \wedge P\beta_k') \in \Gamma_{q_j}$

Since $\Gamma_{q^{**}} < \Gamma_{q_j} < \Gamma_{q''}$, it follows that $F\beta_1, ..., F\beta_n, P\beta_1', ..., P\beta_k' \in \Gamma_{q_j}$. Furthermore, since $\Box\psi, \Box\psi' \in \Gamma_{q_j}$ and $P\varphi \to \Box P\varphi, F\varphi \to \Box F\varphi$, $\Box\varphi \to \Box\Box\varphi$ are axioms, it follows that:

$$\Box(\Box\psi \wedge \Box\psi' \wedge F\beta_1 \wedge ... \wedge F\beta_n \wedge P\beta_1' \wedge ... \wedge P\beta_k') \in \Gamma_{q_j}$$

It is easy to show that $\vdash_{\mathbf{S4}} \Box\gamma \wedge \Diamond\varphi \to \Diamond(\gamma \wedge \varphi)$, and by substituting

$$\Box\psi \wedge \Box\psi' \wedge F\beta_1 \wedge \ldots \wedge F\beta_n \wedge P\beta'_1 \wedge \ldots \wedge P\beta'_k$$

for γ we get:

$$\vdash_{\mathbf{Q_{\Box FP}}} \Box(\Box\psi \wedge \Box\psi' \wedge F\beta_1 \wedge \ldots \wedge F\beta_n \wedge P\beta'_1 \wedge \ldots \wedge P\beta'_k) \wedge \Diamond\varphi \to$$
$$\Diamond(\varphi \wedge \Box\psi \wedge \Box\psi' \wedge F\beta_1 \wedge \ldots \wedge F\beta_n \wedge P\beta'_1 \wedge \ldots \wedge P\beta'_k).$$

So

$$\Diamond(\varphi \wedge \Box\psi \wedge \Box\psi' \wedge F\beta_1 \wedge \ldots \wedge F\beta_n \wedge P\beta'_1 \wedge \ldots \wedge P\beta'_k) \in \Gamma_{q_j}$$

$\Diamond\gamma \to \gamma \vee F\gamma \vee P\gamma$ is an axiom, so:

- $(\varphi \wedge \Box\psi \wedge \Box\psi' \wedge F\beta_1 \wedge \ldots \wedge F\beta_n \wedge P\beta'_1 \wedge \ldots \wedge P\beta'_k) \in \Gamma_{q_j}$
  or
- $F(\varphi \wedge \Box\psi \wedge \Box\psi' \wedge F\beta_1 \wedge \ldots \wedge F\beta_n \wedge P\beta'_1 \wedge \ldots \wedge P\beta'_k) \in \Gamma_{q_j}$
  or
- $P(\varphi \wedge \Box\psi \wedge \Box\psi' \wedge F\beta_1 \wedge \ldots \wedge F\beta_n \wedge P\beta'_1 \wedge \ldots \wedge P\beta'_k) \in \Gamma_{q_j}$

This contradicts $(**)$ above. □

To finish off the construction, let $\langle Q, < \rangle = \bigcup\{\langle Q_n, <_n \rangle \mid n \in \mathbb{N}\}$. It is not difficult to check that $\langle Q, < \rangle$ is a countable dense unbounded linear order (the formula $\Box\top$ ensures both unboundedness and density), so we will move directly on to showing that $\langle Q, \tau \rangle$ satisfies $\Delta \cup \{\neg\varphi\}$. Again let $\langle Q, \tau \rangle$ have the Henkin valuation $v(p) = \{q \mid p \in \Gamma_q\}$.

Proposition 3.19. $\langle Q, \tau, v \rangle, q \models \varphi \Leftrightarrow \varphi \in \Gamma_q.$

Proof.   By induction on the complexity of φ. We will treat the cases $\varphi = \Box\psi$ and $\varphi = F\psi$.

$\varphi = \Box\psi$

[$\Rightarrow$] Assume $\Diamond\neg\psi \in \Gamma_q$. Let U' be an arbitrary open containing q and let q^* and q' be points in U' such that $q^* < q < q'$. Then there is a stage j in the construction such that $q, q^*, q' \in Q_{j-1}$ and $\Diamond\neg\psi$ is being treated. When stage j is completed, there is a point $x \in Q_j$ such that $q^* < x < q'$ and $\neg\psi \in \Gamma_x$. It then follows from the inductive hypothesis that $x \models \neg\psi$. So $\langle Q, \tau, v \rangle, q \models \Diamond\neg\psi$.

[$\Leftarrow$] Assume $\Box\psi \in \Gamma_q$. We know at some stage j and step i such there are points q^{**} and q'' immediately before and after q in the $<^i_j$ ordering such that $\Gamma_{q^{**}} <_{\Box\psi} \Gamma_q <_{\Box\psi} \Gamma_{q''}$. We claim that for all $r, s, t \in Q^m_p$,

$m > i$, $p \geq j$, if $q^{**} \leq r < s < t \leq q''$, then $\Gamma_r <_{\square\psi} \Gamma_s <_{\square\psi} \Gamma_t$. We will argue by induction on the steps of the construction.

Assume at step Q_p^{m+1}, $m \geq i$, $p \geq j$ a point q' is added between q^{**} and q''. By inspecting the construction, it is clear that q' is added immediately before or after the point $q_m \in Q_p$. Without loss of generality, assume it is placed immediately after q_m. By Proposition 3.14 $<_{\square\psi}$ is transitive, so it suffices to show that $\Gamma_{q_m} <_{\square\psi} \Gamma_{q'} <_{\square\psi} \Gamma_{q_v}$, where q_v is the immediate successor of q_m in the $<_p^m$ ordering. We will only check that $\Gamma_{q_m} <_{\square\psi} \Gamma_{q'}$, showing $\Gamma_{q'} <_{\square\psi} \Gamma_{q_v}$ is similar.

By Condition *(ii)* on stages and steps, we know that $\Gamma_{q_m} < \Gamma_{q'} < \Gamma_{q_v}$. Furthermore, by the inductive hypothesis we know that $\Gamma_{q_m} <_{\square\psi} \Gamma_{q_v}$. By inspecting each stage of the construction, we see that $\{\square\varphi \mid \Gamma_{q_m} <_{\square\varphi} \Gamma_{q_v}\} \subseteq \Gamma_{q'}$, so $\square\psi \in \Gamma_{q'}$. Now assume $H\varphi \in \Gamma_{q_m}$. As $\Gamma_{q_m} <_{\square\psi} \Gamma_{q_v}$, it follows that $H(\varphi \vee \square\psi) \in \Gamma_{q_v}$. It is easy to check that this entails $HH(\varphi \vee \square\psi)$ is also in Γ_{q_v}. So by the fact that $\Gamma_{q'} < \Gamma_{q_v}$, it follows that $H(\varphi \vee \square\psi) \in \Gamma_{q'}$. Showing $G\varphi \in \Gamma_{q'} \Rightarrow G(\varphi \vee \square\psi) \in \Gamma_{q_m}$ is similar.

From the above, it follows that for all r, s, t such that $q^{**} \leq r < s < t \leq q''$, $\Gamma_r <_{\square\psi} \Gamma_s <_{\square\psi} \Gamma_t$. This entails that for all q' such that $q^{**} < q' < q''$, $\square\psi \in \Gamma_{q'}$. And since $\square\psi \to \psi$ is an axiom, it follows that ψ is also in all $\Gamma_{q'}$. Then by the inductive hypothesis on the complexity of formulas, $q' \models \psi$ for all q', $q^{**} < q' < q''$. By letting U be the open $U = \{x \mid q^{**} < x < q''\}$, we get that $\langle Q, \tau, v\rangle, q \models \square\psi$.

$\varphi = F\psi$

$[\Rightarrow]$ Assume $F\psi \in \Gamma_q$. Then there is some stage j such that $q \in Q_{j-1}$ and $F\psi$ is being treated at j. After this stage, there is a point $q' > q$ such that $\psi \in \Gamma_{q'}$. Then by the inductive hypothesis $q' \models \psi$, so $\langle Q, \tau, v\rangle, q \models F\psi$.

$[\Leftarrow]$ Assume $\langle Q, \tau, v\rangle, q \models F\psi$. Then there is some $q' > q$ such that $q' \models \psi$ and by the inductive hypothesis $\psi \in \Gamma_{q'}$. Since q and q' are related by $<$ it follows that $F\psi \in \Gamma_q$. $\qquad\square$

It follows as before that $\mathbf{Q_{\square FP}}$ is strongly complete with respect to $\langle Q, \tau\rangle$.

3.5 Extensions to $\mathbb{R}$: The Temporal Case

So far we have seen the logics **S4** and $\mathbf{Q_{\square FP}}$ are strongly complete with respect to the rational line. While this is interesting, $\mathbb{Q}$ is not a particularly rich topological structure. From a mathematical perspective, $\mathbb{R}$ is sometimes viewed as more interesting. And indeed extensions to $\mathbb{R}$ have been considered for the logics $\mathbf{Q_{FP}}$, $\mathbf{S4}$, and $\mathbf{Q_{\Box FP}}$. Working indirectly through Kripke frames, the corresponding problems are found to be technical, difficult, and open, respectively.

As mentioned before, one impressive feature of the temporal construction method is the ease in which completeness on $\langle \mathbb{Q}, < \rangle$ is extended to completeness on $\langle \mathbb{R}, < \rangle$. Thus when attempting topological extensions of $\langle \mathbb{Q}, \tau \rangle$ to $\langle \mathbb{R}, \tau \rangle$ for $\mathcal{L}$ and $\mathcal{L}^{\Box FP}$ using the construction method, it is useful to have the original temporal proof in mind. The proof will make use of the following well known facts.

Fact 3.20. $\langle \mathbb{R}, < \rangle$ *is the unique complete linear ordering that has a countable dense subset isomorphic to* $\langle \mathbb{Q}, < \rangle$.

Fact 3.21. *If* $\langle W, < \rangle$ *is a dense unbounded linearly ordered set, then there exists a continuous unbounded linearly ordered set* $\langle W' <' \rangle$ *such that:*

$$W \subset W' \text{ and } < \text{ and } <' \text{ agree on } W$$
$$W \text{ is dense in } W'$$

It follows from the above that the dense, unbounded, linear order $\langle Q, < \rangle$ constructed in **Section** 2.7 can be extended to a frame $\langle R, < \rangle \cong \langle \mathbb{R}, < \rangle$. Using our intuitions from first order logic as a guide, after extending $\langle Q, < \rangle$ to $\langle R, < \rangle$ we should associate every new point with a $\mathbf{Q_{FP}}$ maximally consistent set that preserves the $<$ ordering. Then we should extend our Henkin valuation to include these new points, and check that our new model $\langle R, < \rangle$ still satisfies $\Delta \cup \{\neg \chi\}$. However, in this case, our first order intuitions fail us. This is one of the rare practical occurrences where the basic modal language is, in some respects, more expressive than its first order counterpart.

Theorem 3.22. Every dense unbounded linear order is elementary equivalent.

Let **DULO** be the first order theory containing axioms for density, unboundedness, and linearity. Then by the above classical theorem:

$$\langle \mathbb{R}, < \rangle \models \varphi \Leftrightarrow \vdash_{\mathbf{DULO}} \varphi \Leftrightarrow \langle \mathbb{Q}, < \rangle \models \varphi$$

It follows immediately that continuity is not first order definable. However, continuity is definable in $\mathcal{L}^{FP}$. Let $A\psi$ denote $\psi \wedge G\psi \wedge H\psi$.

Proposition 3.23. $\langle W, < \rangle \models A(G\varphi \rightarrow PG\varphi) \rightarrow (G\varphi \rightarrow H\varphi) \Leftrightarrow$ $\langle W, < \rangle$ is continuous.

> **Proof.** $[\Rightarrow]$ Assume $\langle W, < \rangle$ is not continuous. Then there is some bounded set $X \subseteq W$ such that $\sup X \notin W$. Let $\langle W, < \rangle$ have the valuation $v(p) = W/X$. Then for all y such that $y < \sup X$, $y \models F\neg p$

and for all y such that $y > \sup X$, $y \models Gp \wedge PGp$. Let z be a point greater than $\sup X$. Then $z \models A(Gp \rightarrow PGp)$ and $z \not\models Gp \rightarrow Hp$.

[$\Leftarrow$] Assume $\langle W, < \rangle$ is continuous and $\langle W, <, v \rangle, w \models A(G\varphi \rightarrow PG\varphi)$. Furthermore, assume towards a contradiction that $w \models G\varphi \wedge P\neg\varphi$. Then the set $X = \{x \mid x < w \text{ and } x \models \neg\varphi\}$ is nonempty. Since W is continuous, $\sup X = w' \in W$. So $w' \models G\varphi \wedge \neg PG\varphi$. But since $w' \leq w$, this means $w \models P\neg(G\varphi \rightarrow PG\varphi) \vee \neg(G\varphi \rightarrow PG\varphi)$. So $w \not\models A(G\varphi \rightarrow PG\varphi)$, contrary to assumption. $\qquad\square$

It should be clear that with continuity we are already getting more expressive power than expected out of the $\mathcal{L}^{FP}$. Intuitively, we would strongly guess no additional properties of $\langle \mathbb{R}, < \rangle$ can be expressed in the language. However, we still want to prove this fact. Let $\mathbf{R_{FP}}$ be the logic $\mathbf{Q_{FP}} + A(G\varphi \rightarrow PG\varphi) \rightarrow (G\varphi \rightarrow H\varphi)$. We again follow an argumennt of de Jongh and Veltman [7].

Theorem 3.24. $\mathbf{R_{FP}}$ is strongly complete with respect to $\langle \mathbb{R}, < \rangle$.

Proof.   Assume $\Delta \not\vdash_{\mathbf{R_{FP}}} \chi$. First carry out the temporal construction procedure for $\langle Q, < \rangle$ as in **Section 2.7**, except this time replacing $\mathbf{Q_{FP}}$ maximally consistent sets with $\mathbf{R_{FP}}$ ones. The end result of the construction will be a countable, dense, unbounded linear order $\langle Q, < \rangle$ which satisfies $\Delta \cup \{\neg\chi\}$. Now extend $\langle Q, < \rangle$ to a structure $\langle R, < \rangle \cong \langle \mathbb{R}, < \rangle$. We know many new points will be added by this procedure, and we want to associate each of them with a $\mathbf{R_{FP}}$ maximally consistent set such that $\langle R, < \rangle$ still satisfies $\Delta \cup \{\neg\chi\}$.

An obvious candidate for the irrational newcomer $x \in R \backslash Q$ is Γ_x, an $\mathbf{R_{FP}}$ maximally consistent extension of:

$$\Gamma = \{\alpha \mid \exists q \in Q \text{ such that } q < x \text{ and } G\alpha \in \Gamma_q\} \cup$$
$$\{\beta \mid \exists q' \in Q \text{ such that } x < q' \text{ and } H\beta \in \Gamma_{q'}\}$$

In this case, the obvious candidate turns out to be the correct one. $\qquad\square$

Lemma 3.25. Γ is consistent.

Proof.   Assume not. Then there are $q_1, ..., q_m < x$ with $G\alpha_i \in \Gamma_{q_i}$ for $1 \leq i \leq m$ and $q'_1, ..., q'_n > x$ with $H\beta_j \in \Gamma_{q'_j}$ for $1 \leq j \leq n$ such that:

$$\vdash_{\mathbf{R_{FP}}} \neg(\alpha_1 \wedge ... \wedge \alpha_m \wedge \beta_1 \wedge ... \wedge \beta_n)$$

However, since $\langle Q, < \rangle$ is dense, there must be some q^* such that $q_1, ..., q_m < q^* < q'_1, ..., q'_n$ and since the $<$ ordering on $\langle Q, < \rangle$ respects $<$, $\alpha_1 \wedge ... \alpha_m \wedge \beta_1 \wedge ... \wedge \beta_n \in \Gamma_{q^*}$. □

So we only need to check that extending the Henkin valuation to include the added irrational points preserves the desired harmony between the syntactic and the semantic.

Lemma 3.26. For all $r \in R\backslash Q$, if $F\psi \in \Gamma_r$ then there is some $q \in Q, r < q$, such that $\psi \in \Gamma_q$.

> **Proof.** Assume $F\psi \in \Gamma_r$, $r \in R\backslash Q$, and for all $q'' \in Q$ such that $r < q''$, $\neg\psi \in \Gamma_{q''}$. Then we know from the properties of our constructed model $\langle Q, <, \nu \rangle$ that for all $q'' \in Q$ such that $r < q''$, $G\neg\psi \wedge PG\neg\psi \in \Gamma_{q''}$. Furthermore, we also know that for all $q^* \in Q$ such that $q^* < r$, $F\psi \in \Gamma_{q^*}$, for otherwise $G\neg\psi \in \Gamma_r$ by the definition of Γ_r. So for every $q \in Q$, either $PG\neg\psi \in \Gamma_q$ or $F\psi \in \Gamma_q$. It follows that for all $q \in Q$, $A(G\neg\psi \to PG\neg\psi) \in \Gamma_q$. Now let q' be a point in Q such that $q < q'$. From $A(G\neg\psi \to PG\neg\psi) \wedge G\neg\psi \in \Gamma_{q'}$ and the continuity axiom it follows that $H\neg\psi \in \Gamma_{q'}$. But this contradicts the fact that $F\psi \in \Gamma_{q^*}$ for all $q^* < r$, and $\langle Q, <, \nu \rangle$ agrees with the $<$ ordering. □

Lemma 3.27. $\langle R, <, \nu \rangle \models \varphi \Leftrightarrow \varphi \in \Gamma_r$.

> **Proof.** By induction on the complexity of φ. We will only treat the case $\varphi = F\psi$.
>
> [$\Leftarrow$] Assume $F\psi \in \Gamma_r$. Then by the above lemma and the $<$ ordering on $\langle Q, < \rangle$, there is some $q \in Q$ such that $r < q$ and $\psi \in \Gamma_q$. By induction hypothesis, $q \models \psi$, so $\langle R, <, \nu \rangle, r \models F\psi$.
>
> [$\Rightarrow$] Assume $F\psi \notin \Gamma_r$. It is not hard to check that $\neg\psi \in \Gamma_q$ for all $q \in Q$ such that $r < q$. So assume towards a contradiction that there is some $r' \in R\backslash Q$ such that $r < r'$ and $\psi \in \Gamma_{r'}$. Since $\langle Q, < \rangle$ is a dense subset of $\langle R, < \rangle$, there exists a point $q' \in Q$ such that $r < q' < r'$. However, we know $G\neg\psi \in \Gamma_{q'}$ and thus by the definition of $\Gamma_{r'}$ that $\neg\psi \in \Gamma_{r'}$. This contradicts our assumption. So for all $r'' \in R$ such that $r < r''$, $\neg\psi \in \Gamma_{r''}$. By the inductive hypothesis all $r'' \models \neg\psi$ and $\langle R, < \nu \rangle, r \not\models F\psi$. □

Since $\Delta \cup \{\neg\chi\} \subseteq \Gamma_{q^\#}$, it readily follows that $\mathbf{R_{FP}}$ is strongly complete with respect to $\langle \mathbb{R}, < \rangle$.

We have now seen that our constructed $\mathcal{L}^{FP}$ model $\langle Q, <, v\rangle$ can be extended to a new model $\langle \mathbb{R}, <, v\rangle$ which preserves the truth of $\mathcal{L}^{FP}$ formulas on the original model. Now let us consider this procedure for $\langle Q, \tau, v\rangle$ and $\langle \mathbb{R}, \tau, v\rangle$ in the languages $\mathcal{L}$ and $\mathcal{L}^{\Box FP}$.

Let us first consider the extension of $\langle Q, \tau, v\rangle$ to $\langle \mathbb{R}, \tau, v\rangle$ in $\mathcal{L}$. Unlike the temporal case, we know in advance that **S4** is the logic of both $\langle Q, \tau\rangle$ and $\langle \mathbb{R}, \tau\rangle$, so we will only need to fill in the missing irrational points and associate them with **S4** maximally sets. In all other respects, let us proceed as above.

Assume $\Delta \nvdash_{\mathbf{S4}} \chi$ and let $\langle Q, \tau, v\rangle$ be our constructed topological model which satisfies $\Delta \cup \{\neg\chi\}$. Extend $\langle Q, <\rangle$ to a structure $\langle R, <\rangle \cong \langle \mathbb{R}, <\rangle$. Now associate each $x \in R\backslash Q$ with Γ_x, an **S4** maximally consistent extension of:

$$\Gamma = \{\Box\psi \mid \exists q_1, q_2 \in Q \text{ such that } q_1 < x < q_2 \text{ and } \forall q \in Q \text{ such that}$$
$$q_1 < q < q_2, \psi \in \Gamma_q\} \cup \{\Diamond\varphi \mid \forall q_3, q_4 \in Q \text{ such that } q_3 < x < q_4, \exists q' \in Q$$
$$\text{such that } q_3 < q' < q_4 \text{ and } \varphi \in \Gamma_{q'}\}$$

Lemma 3.28. $\langle R, \tau, v\rangle, r \models \varphi \Leftrightarrow \varphi \in \Gamma_r$.

> **Proof.**   By induction on the complexity of φ. We will only treat the case $\varphi = \Box\psi$.
>
> $[\Leftarrow]$ Assume $\Box\psi \in \Gamma_r$. We claim that $\exists q_1, q_2 \in Q$ such that $q_1 < r < q_2$ and $\forall q \in Q, q_1 < q < q_2, \psi \in \Gamma_q$. If $r \in Q$, this holds by the construction of $\langle Q, \tau\rangle$. And if $r \in R\backslash Q$ this follows from the definition of Γ_r (since $\Diamond\neg\psi \notin \Gamma_r$).
>
> Now assume towards a contradiction that there is some $r' \in R\backslash Q$ such that $q_1 < r' < q_2$ and $\neg\psi \in \Gamma_{r'}$. Then by maximal consistency $\Box\psi \notin \Gamma_{r'}$. It follows from the definition of $\Gamma_{r'}$ that for all $q_3, q_4 \in Q$ such that $q_3 < r' < q_4$, there exists some $q' \in Q$, $q_3 < q' < q_4$, and $\neg\psi \in \Gamma_{q'}$. But this means there is a $q' \in Q$ such that $q_1 < q' < q_2$ and $\neg\psi \in \Gamma_{q'}$, contradicting the above.
>
> So for all $x \in R$ such that $q_1 < x < q_2, \psi \in \Gamma_x$. By letting $U = \{y \mid q_1 < x < q_2\}$ and the inductive hypothesis we get that $\langle R, \tau, v\rangle, r \models \Box\psi$.
>
> $[\Rightarrow]$ Assume $\Box\psi \notin \Gamma_r$. We claim that $\forall q_3, q_4 \in Q$ such that $q_3 < r < q_4, \exists q' \in Q, q_3 < q' < q_4$, and $\neg\psi \in \Gamma_{q'}$. If $r \in Q$ this holds by the construction of $\langle Q, <\rangle$ and if $r \in R\backslash Q$ then this holds by definition of Γ_r (since $\Box\psi \notin \Gamma_r$). So it follows from the inductive hypothesis that $\langle R, \tau, v\rangle, r \nvDash \Box\psi$. $\square$

In order to show our constructed model $\langle Q, \tau, v \rangle$ can be extended to $\langle \mathbb{R}, \tau, v \rangle$ while preserving the truth of $\mathcal{L}$ formulas, it only remains to be shown that Γ is consistent. Unfortunately, this cannot be done. For unlike the previous cases, our natural syntactic candidate to associate with a new point may be inconsistent. For example, assume $x \in R \backslash Q$ and $\langle Q, \tau \rangle$ has the valuation $v(p) = \{q \mid q < x\}$. Then according to the definition of Γ_x, $\Diamond p$ and $\Diamond \neg p$ are in Γ_x. Furthermore, for every $q \in Q$, either $q \models \Box p$ or $q \models \Box \neg p$, so $\Box(\Box p \vee \Box \neg p) \in \Gamma_x$. But $\vdash_{\mathbf{S4}} \Box(\Box p \vee \Box \neg p) \rightarrow \neg(\Diamond p \wedge \Diamond \neg p)$ and this entails that $\neg(\Diamond p \wedge \Diamond \neg p) \wedge (\Diamond p \wedge \Diamond \neg p) \in \Gamma_x$.

So already things are worse than the temporal case, but one might hope that they are still somewhat better than the model-theoretic counterpart. As mentioned before, the completeness of $\mathbf{S4}$ with respect to $\langle \mathbb{R}, \tau \rangle$ can be established model-theoretically, but with tricky diversions through finite transitive and reflexive trees and subsets of the Cantor Space. As we have seen, using the construction method we have some control over how formulas are satisfied on $\langle Q, \tau \rangle$ and perhaps we can make use this control when jumping to $\langle \mathbb{R}, \tau \rangle$.

Looking at the failure of Γ, it is clear what property we want our constructed model $\langle Q, \tau, v \rangle$ to have: if there are rational $\varphi_1, .., \varphi_n$ sequences approaching an irrational point x, then there is a rational sequence $\varphi_1 \wedge \dots \wedge \varphi_n$ approaching x. We could then show Γ is consistent. For assume not. Then there are formulas $\Box \psi, \Diamond \varphi_1, \dots, \Diamond \varphi_n$ such that

- $\exists q_1, q_2 \in Q, q_1 < x < q_2$ and for all $q' \in Q, q_1 < q' < q_2, \psi \in \Gamma_{q'}$
- $\forall q_3, q_4 \in Q, q_3 < x < q_4$, there exists some $q_i' \in Q$ such that $q_3 < q_i' < q_4$ and $\varphi_i \in \Gamma_{q_i'}$, for $1 \leq i \leq n$
- $\vdash_{\mathbf{S4}} \Box \psi \rightarrow \neg(\Diamond \varphi_1 \wedge \dots \wedge \Diamond \varphi_n)$

But as there are rational sequences $\varphi_1, \dots, \varphi_n$ approaching x, by assumption there is a rational sequence $\Diamond(\varphi_1 \wedge \dots \wedge \varphi_n)$ approaching x. This implies there is some rational point q' in between q_1 and q_2 such that: $\Box \psi, \Diamond \varphi_1, \dots, \Diamond \varphi_n \in \Gamma_{q'}$.

However, after some reflection, it is clear that the desired property of $\langle Q, \tau, v \rangle$ is (at least) very difficult to ensure. For while we have control on how formulas are satisfied on the rationals, this tells us very little about how they will line up on the irrationals. There is no natural way to force a construction of $\langle Q, \tau, v \rangle$ to tell us what kinds of sequences are being created towards irrational points.

The underlying difficulty of extending $\langle Q, \tau, v \rangle$ to $\langle \mathbb{R}, \tau, v \rangle$ in $\mathcal{L}$ is that there is no way of expressing any topological relation between the points in the language. We can see this defect clearly when comparing our failed attempt in $\langle \mathbb{R}, \tau \rangle$ to the successful temporal proof for $\langle \mathbb{R}, < \rangle$. In the temporal construction, each state of affairs expressible in $\mathbf{R_{FP}}$ (i.e. each maximal consistent set) is related to every other state of affairs in a way that respects order structure on $\langle \mathbb{Q}, < \rangle$. Furthermore, each description of an $\mathbf{R_{FP}}$ state of affairs existing in

$\langle \mathbb{Q}, < \rangle$ says that if there is a bounded φ sequence, then this φ sequence has a least upper bound respecting the $<$ ordering. Thus when extending $\langle \mathbb{Q}, < \rangle$ to $\langle \mathbb{R}, < \rangle$, we know we can associate an irrational point x with its natural syntactic candidate Γ_x, since this state of affairs has already been described to exist in $\langle \mathbb{Q}, < \rangle$.

The temporal construction for $\langle \mathbb{R}, < \rangle$ goes so smoothly because there is a kind of expressive harmony between what can be said to occur in the language and the structure $\langle \mathbb{R}, < \rangle$. Clearly this is not the case in $\langle \mathbb{R}, \tau \rangle$, where we are working with a fairly rich topological structure and a language that can express almost no topological properties of the structure. In this sense, it is reasonable to think that the seemingly more difficult construction of $\langle \mathbb{R}, \tau \rangle$ in $\mathcal{L}^{\Box FP}$ would fare better, since additional topological properties such as connectedness are expressible in $\mathcal{L}^{\Box FP}$. However, $\mathcal{L}^{\Box FP}$ is still too weak of a language for things to go smoothly. In order to successfully extend $\langle Q, \tau \rangle$ to $\langle \mathbb{R}, \tau \rangle$ we need to be able to express in the language when a particular $\Box$ stretch ends, which is a much stronger property than connectedness.

3.7 $\mathbf{Q_{SU}}$ on $\mathbb{Q}$

Perhaps at some point in the preceding discussion, the reader made notice of the resemblance between $\mathcal{L}^{\Box FP}$ and the popular temporal Since and Until language $\mathcal{L}^{SU}$. Indeed, $\mathcal{L}^{SU}$ has all the expressive power of $\mathcal{L}^{\Box FP}$, plus the additional property we desired in our attempted constructions of $\langle \mathbb{R}, \tau \rangle$: the ability to express up to what point a $\Box$ stretch holds. In this section we will examine how this additional property simplifies completeness proofs using the construction method.

The Since/Until language $\mathcal{L}^{SU}$ contains two binary operators U and S, and is built up in the usual way from a set of propositional letters and these two operators. Intuitively, the formulas $U(\varphi, \psi)$ and $S(\varphi, \psi)$ are supposed to mean "until φ is true, ψ holds" and "since φ was true, ψ has held." More formally, given a model $M = \langle W, <, v \rangle$, where $<$ is a partial order, we define when an $\mathcal{L}^{SU}$ formula γ is true at a point w as follows:

- $w \models U(\varphi, \psi)$ iff $\exists w'' > w$ such that $w'' \models \varphi$ and $\forall w', w < w' < w''$, $w' \models \psi$
- $w \models S(\varphi, \psi)$ iff $\exists w'' < w$ such that $w'' \models \varphi$ and $\forall w', w > w' > w''$, $w' \models \psi$

As the reader may have noticed, $\mathcal{L}^{SU}$ is expressively complete over $\mathcal{L}^{\Box FP}$. We can define $\Box$, F and P as follows:

$$F\psi := U(\psi, \top)$$
$$P\psi := S(\psi, \top)$$

$$\Box\psi := S(\top, \psi) \wedge \psi \wedge U(\top, \psi)$$

However, in $\mathcal{L}^{SU}$ it is also possible to say additional things like "until we are in a $\Box\varphi$ stretch, we are in a $\Box\psi$ stretch" $(U(\Box\varphi, \Box\psi))$.

The language $\mathcal{L}^{SU}$ was introduced in Kamp [9] and, as may be guessed from its name, the original interest was temporal rather than topological. After the first most general completeness results for partial orders were established, attention naturally turned to mathematical structures such as $\langle \mathbb{N}, < \rangle$, $\langle \mathbb{Q}, < \rangle$ and $\langle \mathbb{R}, < \rangle$. Kamp made progress in these problems but never published his results, as his proofs were regarded as very unwieldy. The first published results in the area are found in Burgess [6], where axiomatizations of $\langle \mathbb{N}, < \rangle$ and $\langle \mathbb{Q}, < \rangle$ are given. However, even Burgess's proofs are difficult, and make crucial use of high-powered technical results. Let us see how the construction method can be used to both simplify matters and establish strong completeness.

Let $\mathbf{Q_{SU}}$ be the logic containing axioms:

(1a)	$G(\varphi \rightarrow \psi) \rightarrow (U(\varphi, \gamma) \rightarrow U(\psi, \gamma))$
(2a)	$G(\varphi \rightarrow \psi) \rightarrow (U(\gamma, \varphi) \rightarrow U(\gamma, \psi))$
(3a)	$\varphi \wedge U(\psi, \gamma) \rightarrow U(\psi \wedge S(\varphi, \gamma), \gamma)$
(4a)	$U(\varphi, \psi) \wedge \neg U(\gamma, \psi) \rightarrow U(\varphi \wedge \neg\gamma, \psi)$
(5a)	$U(\varphi, \psi) \wedge \neg U(\varphi, \gamma) \rightarrow U(\psi \wedge \neg\gamma, \psi)$
(6a)	$U(\varphi, \psi) \rightarrow U(\varphi, \psi \wedge U(\varphi, \psi))$
(7a)	$U(\psi \wedge U(\varphi, \psi), \psi) \rightarrow U(\varphi, \psi)$
(8a)	$U(\varphi, \psi) \wedge U(\gamma, \zeta) \rightarrow U(\gamma \wedge \psi \wedge U(\varphi, \psi), \psi \wedge \zeta) \vee$
	$U(\varphi \wedge \gamma, \psi \wedge \zeta) \vee U(\varphi \wedge \zeta \wedge U(\gamma, \zeta), \psi \wedge \zeta)$
(9a)	$U(\top, \top)$
(10)	$\neg U(\top, \bot)$

and *(1b)–(9b)*, where all occurrences of S are replaced by U in *(1a)–(9a)*, and vice versa.

Theorem 3.29. $\mathbf{Q_{SU}}$ is strongly complete with respect to $\langle \mathbb{Q}, < \rangle$.

Proof. Assume $\Delta \nvdash_{\mathbf{Q_{SU}}} \chi$. We will construct a frame $\langle Q, < \rangle \cong \langle \mathbb{Q}, < \rangle$ in stages which satisfies $\Delta \cup \{\neg\chi\}$.

A stage n consists of:

(i) a finite set $Q_n = \{q_0, q_1 ..., q_k\}$
(ii) a linear ordering $<_n$ on Q_n
(iii) an assignment of a $\mathbf{Q_{SU}}$ maximally consistent set Γ_{q_i} to each $q_i \in Q_n$

Definition 3.30. We say $\Gamma_q <_\varphi \Gamma_{q'}$ if $\psi \in \Gamma_{q'} \Rightarrow U(\psi, \varphi) \in \Gamma_q$.

Note the simplicity of the $<$ relation compared to the $<_{\Box\psi}$ relation in $\mathcal{L}^{\Box FP}$. By adding slightly more expressive power, we no longer have to force the temporal operators G and H to interact correctly with $\Box$. The language now takes care of this for us. $\qquad\square$

Proposition 3.31. $\qquad \Gamma_q <_\varphi \Gamma_{q'}$ and $\Gamma_q <_\psi \Gamma_{q'} \Rightarrow \Gamma_q <_{\varphi\wedge\psi} \Gamma_{q'}$

Proof. Assume not. Then there is a $\gamma \in \Gamma_{q'}$ such that $U(\gamma,\varphi) \in \Gamma_q$, $U(\gamma,\psi) \in \Gamma_q$ and $\neg U(\gamma, \varphi\wedge\psi) \in \Gamma_q$. By axiom *(8a)*

(a) $\qquad U(\gamma, \varphi\wedge\psi) \in \Gamma_q$ or
(b) $\qquad U(\gamma \wedge \varphi \wedge U(\varphi,\psi), \varphi\wedge\psi) \in \Gamma_q$ or
(c) $\qquad U(\gamma \wedge \psi \wedge U(\gamma,\psi), \varphi\wedge\psi) \in \Gamma_q$

(a) contradicts the assumption that $\neg U(\gamma, \varphi\wedge\psi) \in \Gamma_q$ straight away, so assume either *(b)* or *(c)* hold instead. It is a basic fact that $\vdash_{\mathbf{QSU}} U(\alpha \wedge \beta, \chi) \leftrightarrow U(\alpha, \chi) \wedge U(\beta, \chi)$, so it follows again that $U(\gamma, \varphi \wedge \psi) \in \Gamma_q$, contradicting the consistency of Γ_q. $\qquad\square$

Similar to previous constructions, when adding a new point q' in between points q and q'' such that $\Gamma_q <_\varphi \Gamma_{q''}$, we want to take care that $\varphi \in \Gamma_{q'}$ and $\Gamma_q <_\varphi \Gamma_{q'} <_\varphi \Gamma_{q''}$. In other words, we want to be sure that

$$\{\varphi\} \cup \{\neg\psi \mid \neg U(\psi, \varphi) \in \Gamma_q\} \cup \{U(\gamma, \varphi) \mid \gamma \in \Gamma_{q''}\} \subseteq \Gamma_{q'}$$

In order to show that this is possible it is useful to prove the following proposition.

Proposition 3.32. Assume $\Gamma_q <_\varphi \Gamma_{q''}$, $\neg U(\psi, \varphi) \in \Gamma_q$ and $\gamma \in \Gamma_{q''}$. Then $\Gamma_q <_{\varphi\wedge\neg\psi\wedge U(\gamma,\varphi)} \Gamma_{q''}$.

Proof. Assume $\alpha \in \Gamma_{q''}$. We want to show $U(\alpha, \varphi \wedge \neg\psi \wedge U(\gamma,\varphi)) \in \Gamma_q$.

Since $\Gamma_q <_\varphi \Gamma_{q''}$, we know that $U(\alpha, \varphi) \in \Gamma_q$.

Now we will show that $U(\alpha, \neg\psi) \in \Gamma_q$. Assume towards a contradiction that $\neg U(\alpha, \neg\psi) \in \Gamma_q$. From $\alpha \in \Gamma_{q''}$ and $\Gamma_q <_\varphi \Gamma_{q''}$, we know that $U(\alpha, \varphi) \in \Gamma_q$. Then it follows from axiom 5a) that $U(\varphi \wedge \psi, \varphi) \in \Gamma_q$. But this implies $U(\psi, \varphi) \in \Gamma_q$, contradicting our assumption.

Finally we will show $U(\alpha, U(\gamma, \varphi)) \in \Gamma_q$. Assume towards a contradiction that $\neg U(\alpha, U(\gamma, \varphi)) \in \Gamma_q$. From $\alpha, \gamma \in \Gamma_{q''}$ and $\Gamma_q <_\varphi \Gamma_{q''}$, we know that $U(\alpha \wedge \gamma, \varphi) \in \Gamma_q$. Then from axiom *(6a)* it follows that

$U(\alpha \wedge \gamma, \varphi \wedge U(\alpha \wedge \gamma, \varphi)) \in \Gamma_q$. But this implies that $U(\alpha, U(\gamma, \varphi)) \in \Gamma_q$, again contradicting the consistency of Γ_q.

It follows from the above and **Proposition 3.31** that $U(\alpha, \varphi \wedge \neg \psi \wedge U(\gamma, \varphi)) \in \Gamma_q$. □

We are now ready to begin the construction of $\langle Q, < \rangle$. Let $\varphi_0 \varphi_1 \varphi_2 ...$ be an enumeration of all U, $\neg U$, S and $\neg S$ formulas, in which each formula is repeated infinitely many times.

Stage 0: Let $Q_0 = \{q_0\}$ and associate q_0 with Γ_{q_0}, a $\mathbf{Q_{SU}}$ maximally consistent extension of $\Delta \cup \{\neg \chi\}$.

Stage 4n+1: Let $\neg U(\varphi, \psi)$ be the next $\neg U$ formula in the enumeration. For all $q \in Q_{4n}$, if

(a) $\neg U(\varphi, \psi) \in \Gamma_q$ and
(b)   there is a $u' > q$ such that $\varphi \in \Gamma_{u'}$ and
(c)   for all $t \in Q_{4n}$ such that $q < t < u'$, $\psi \in \Gamma_t$

then let u be the least point $q < u$ with $\varphi \in \Gamma_u$ and let r be the greatest point $q \leq r < u$ such that $\neg U(\varphi, \psi) \in \Gamma_r$. Let s be the immediate successor of r in $<_{4n}$. Add a new point q' in between r and s and associate q' with $\Gamma_{q'}$, a $\mathbf{Q_{SU}}$ maximally consistent extension of:

$$\Gamma = \{\neg \psi\} \cup \{\gamma \mid \Gamma_r <_\gamma \Gamma_s\} \cup \{\neg \alpha \mid \Gamma_r <_\gamma \Gamma_s \text{ and}$$
$$\neg U(\alpha, \gamma) \in \Gamma_r\} \cup \{U(\beta, \gamma) \mid \Gamma_r <_\gamma \Gamma_s \text{ and } \beta \in \Gamma_s\}$$

Proposition 3.33. Γ is consistent.

> **Proof.**   Assume not. Then there are formulas $\gamma_1, ..., \gamma_k$, $\neg U(\alpha_1, \gamma_1'), ..., \neg U(\alpha_l, \gamma_l'), \gamma_1'', ..., \gamma_m'' \in \Gamma_r$ and $\beta_1, ..., \beta_m \in \Gamma_s$ such that
>
> $\vdash_{\mathbf{Q_{SU}}} \neg(\neg \psi \wedge \gamma_1 \wedge ... \wedge \gamma_k \wedge \neg \alpha_1 \wedge ... \wedge \neg \alpha_l \wedge U(\beta_1, \gamma_1'') \wedge ... \wedge U(\beta_m, \gamma_m''))$
> $\vdash_{\mathbf{Q_{SU}}} G \neg(\neg \psi \wedge \gamma_1 \wedge ... \wedge \gamma_k \wedge \neg \alpha_1 \wedge ... \wedge \neg \alpha_l \wedge U(\beta_1, \gamma_1'') \wedge ... \wedge U(\beta_m, \gamma_m''))$
> $\vdash_{\mathbf{Q_{SU}}} \neg F(\neg \psi \wedge \gamma_1 \wedge ... \wedge \gamma_k \wedge \neg \alpha_1 \wedge ... \wedge \neg \alpha_l \wedge U(\beta_1, \gamma_1'') \wedge ... \wedge U(\beta_m, \gamma_m''))$
>
> From now on let $\zeta = \gamma_1 \wedge ... \wedge \gamma_k \wedge \neg \alpha_1 \wedge ... \wedge \neg \alpha_l \wedge U(\beta_1, \gamma_1'') \wedge ... \wedge U(\beta_m, \gamma_m'')$.

We know from assumptions *(a)-(c)* above that either:

(1) $\Gamma_s = \Gamma_u$ and $\varphi \in \Gamma_s$ or
(2) $\Gamma_s \neq \Gamma_u$ and $\psi \wedge U(\varphi, \psi) \in \Gamma_s$

First assume *(1)* holds. Then since $\varphi \in \Gamma_s$, it follows from **Proposi-tions 3.31** and **3.32** that $U(\varphi, \zeta) \in \Gamma_r$. Furthermore, since $\neg U(\varphi, \psi) \in \Gamma_r$, it follows from axiom *(5a)* that $U(\neg \psi \wedge \zeta, \zeta) \in \Gamma_r$. But this implies $F(\neg \psi \wedge \zeta) \in \Gamma_r$, contradicting the consistency of Γ_r.

So assume *(2)* holds instead. By contraposition of axiom *(7a)*, we have that $\neg U(\varphi, \psi) \rightarrow \neg U(\psi \wedge U(\varphi, \psi), \psi)$. Since $\neg U(\varphi, \psi) \in \Gamma_r$, $\neg U(\psi \wedge U(\varphi, \psi), \psi) \in \Gamma_r$. Furthermore, since $\psi \wedge U(\varphi, \psi) \in \Gamma_s$, by **Propositions 3.31** and **3.32** we have that $U(\psi \wedge U(\varphi, \psi), \zeta) \in \Gamma_r$. So by axiom *(5a)*, $U(\neg \psi \wedge \zeta, \zeta) \in \Gamma_r$. But this implies $F(\neg \psi \wedge \zeta) \in \Gamma_r$, again contradicting the consistency of Γ_r. $\qquad\square$

Stage 4n+2: Let $U(\varphi, \psi)$ be the next U formula in the enumeration. For all $q \in Q_{4n+1}$ if $U(\varphi, \psi) \in \Gamma_q$ and there is no $w > q$ such that:

- $\varphi \in \Gamma_w$
- $\forall r \forall s, q \leq r < s \leq w, \Gamma_r <_\psi \Gamma_s$
- $\forall q', q < q' < w, \psi \in \Gamma_{q'}$

then we need to add a new point somewhere after q such that all of the above properties hold at that point. Let t be the greatest element in the $<_{4n+1}$ ordering, $q \leq t$, such that

- $\forall r \ q \leq r \leq t, U(\varphi, \psi) \in \Gamma_r$
- $\forall r \forall s \ q \leq r < s \leq t, \Gamma_r <_\psi \Gamma_s$
- $\forall r \ q < r \leq t, \psi \in \Gamma_r$

Add a new point u immediately after t. Let v be the immediate successor of t in Q_{4n+1}, provided one exists. Associate u with Γ_u, a $\mathbf{Q_{SU}}$ maximally consistent extension of:

$$\Gamma' = \{\varphi\} \cup \{\neg \alpha \mid \neg U(\alpha, \psi) \in \Gamma_t\} \cup \{\neg \beta \mid \Gamma_t <_\gamma \Gamma_v \text{ and} \\ \neg U(\beta, \gamma) \in \Gamma_t\} \cup \{U(\xi, \gamma) \mid \Gamma_t <_\gamma \Gamma_v \text{ and } \xi \in \Gamma_v\}$$

Proposition 3.34. Γ' is consistent

Proof.   Assume not.   Then there are $\neg U(\alpha_1, \ \psi), ...,$
$\neg U(\alpha_j, \psi), \gamma_1, ..., \gamma_k, \neg U(\beta_1, \ \gamma_1'), ..., \neg U(\beta_l, \ \gamma_l'), \ \gamma_1'', ..., \gamma_m'' \in \Gamma_t$
and $\xi_1, ..., \xi_m \in \Gamma_v$ such that: $\vdash_{\mathbf{Q_{SU}}} \neg(\varphi \wedge \neg \alpha_1 \wedge ... \wedge \neg \alpha_j \wedge \gamma_1 \wedge ... \wedge \gamma_k \wedge \neg \beta_1 \wedge ... \wedge \neg \beta_l \wedge U(\xi_1, \gamma_1'') \wedge ... \wedge U(\xi_m, \gamma_m''))$

$\vdash_{\mathbf{QSU}} G\neg(\varphi \wedge \neg\alpha_1 \wedge ... \wedge \neg\alpha_j \wedge \gamma_1 \wedge ... \wedge \gamma_k \wedge \neg\beta_1 \wedge ... \wedge \neg\beta_l \wedge U(\xi_1, \gamma_1'') \wedge ... \wedge U(\xi_m, \gamma_m''))$

Let $\zeta = \gamma_1 \wedge ... \wedge \gamma_k \wedge \neg\beta_1 \wedge ... \wedge \neg\beta_l \wedge U(\xi_1, \gamma_1'') \wedge ... \wedge U(\xi_m, \gamma_m'')$.

We know that the point t was chosen for one of the following reasons:

(i) $\Gamma_t \not<_\psi \Gamma_v$
(ii) $\Gamma_t <_\psi \Gamma_v$ but $\neg\varphi \wedge \neg U(\varphi, \psi) \in \Gamma_v$
(iii) $\Gamma_t <_\psi \Gamma_v$ but $\neg\varphi \wedge \neg\psi \in \Gamma_v$

First assume *(i)* holds. Then there is some $\chi \in \Gamma_v$ such that $\neg U(\chi, \psi) \in \Gamma_t$. By **Propositions** 3.31 and 3.32, it follows that $U(\chi, \zeta) \in \Gamma_t$ and thus by axiom *(5a)*, we have that $U(\zeta \wedge \neg\psi, \zeta) \in \Gamma_t$. Furthermore, since $U(\varphi, \psi), \neg U(\alpha_1, \psi), ..., \neg U(\alpha_j, \psi) \in \Gamma_t$, by repeated use of axiom *(4a)* we have that $U(\varphi \wedge \alpha_1 \wedge ... \wedge \alpha_n, \psi) \in \Gamma_t$. Putting these facts together, it follows from axiom *(8a)* that:

(a) $U(\zeta \wedge \neg\psi \wedge \psi \wedge U(\varphi \wedge \neg\alpha_1 \wedge ... \wedge \neg\alpha_j, \psi), \psi \wedge \zeta) \in \Gamma_t$ or
(b) $U(\varphi \wedge \neg\alpha_1 \wedge ... \wedge \neg\alpha_j \wedge \zeta \wedge \neg\psi, \psi \wedge \zeta) \in \Gamma_t$ or
(c) $U(\varphi \wedge \neg\alpha_1 \wedge ... \wedge \neg\alpha_j \wedge \zeta \wedge U(\zeta \wedge \neg\psi, \zeta), \psi \wedge \zeta) \in \Gamma_t$

Choice *(a)* straightforwardly contradicts the consistency of Γ_t. However, both *(b)* and *(c)* imply that $F(\varphi \wedge \neg\alpha_1 \wedge ... \wedge \neg\alpha_j \wedge \zeta) \in \Gamma_t$, which also contradicts the consistency of Γ_t.

So assume *(ii)* holds. By **Propositions** 3.31 and 3.32, it follows that $U(\neg\varphi \wedge \neg U(\varphi, \psi), \zeta) \in \Gamma_t$. As before, from $U(\varphi, \psi), \neg U(\alpha_1, \psi), ..., \neg U(\alpha_j, \psi) \in \Gamma_t$ we have $U(\varphi \wedge \neg\alpha_1 \wedge ... \wedge \neg\alpha_j, \psi) \in \Gamma_t$. Then by axiom *(8a)*

(d) $U(\varphi \wedge \neg\alpha_1 \wedge ... \wedge \neg\alpha_j \wedge \zeta \wedge U(\neg\varphi \wedge \neg U(\varphi, \psi), \zeta), \psi \wedge \zeta) \in \Gamma_t$
        or
(e) $U(\varphi \wedge \neg\alpha_1 \wedge ... \wedge \neg\alpha_j \wedge \neg\varphi \wedge \neg U(\varphi\wedge, \psi), \psi \wedge \zeta) \in \Gamma_t$ or
(f) $U(\neg\varphi \wedge \neg U(\varphi, \psi) \wedge \psi \wedge U(\varphi \wedge \neg\alpha_1 \wedge ... \wedge \neg\alpha_j, \psi), \psi \wedge \zeta) \in \Gamma_t$

Clearly *(e)* and *(f)* contradict the consistency of Γ_t. So only *(d)* is possible. However by assumption $F(\varphi \wedge \neg\alpha_1 \wedge ... \wedge \neg\alpha_j \wedge \zeta)$ is inconsistent.

Finally assume *(iii)* holds. It follows from **Propositions** 3.31 and 3.32 that $U(\neg\psi \wedge \neg\varphi, \zeta) \in \Gamma_t$. As above, we have that $U(\varphi \wedge \neg\alpha_1 \wedge ... \wedge \neg\alpha_j, \psi) \in \Gamma_t$. Then it follows from axiom *(8a)* that

(g) $U(\varphi \wedge \neg\alpha_1 \wedge ... \wedge \neg\alpha_j \wedge \neg\psi \wedge \neg\varphi, \psi \wedge \zeta) \in \Gamma_t$ or
(h) $U(\neg\psi \wedge \neg\varphi \wedge \psi \wedge U(\varphi \wedge \neg\alpha_1 \wedge ... \wedge \neg\alpha_j, \psi), \psi \wedge \zeta) \in \Gamma_t$ or

(i) $U(\varphi \wedge \neg \alpha_1 \wedge ... \wedge \neg \alpha_j \wedge \zeta \wedge U(\neg \psi \wedge \neg \varphi, \zeta), \psi \wedge \zeta) \in \Gamma_t$

(g) and *(h)* contradict the consistency of Γ_t directly, and *(i)* implies $F(\varphi \wedge \neg \alpha_1 \wedge ... \wedge \neg \alpha_j \wedge \zeta)$.   □

Definition 3.35. We say $\Gamma_q >_\varphi \Gamma_{q'}$ if $\psi \in \Gamma_q \Rightarrow S(\psi, \varphi) \in \Gamma_{q'}$.

Stages $4n+3$ and $4n+4$ treat the next $\neg S$ and S formulas in the enumeration. They are mirror images of $4n+1$ and $4n+2$, using the $>$ relation in place of $<$.

Proposition 3.36. $\Gamma_q <_\varphi \Gamma_{q'} \Leftrightarrow \Gamma_q >_\varphi \Gamma_{q'}$

> **Proof.**   [$\Rightarrow$] Assume towards a contradiction that $\gamma \in \Gamma_{q'} \Rightarrow U(\gamma, \psi) \in \Gamma_q$ and there is some $\varphi \in \Gamma_q$ such that $\neg S(\varphi, \psi) \in \Gamma_{q'}$. It follows that $\varphi \wedge U(\neg S(\varphi, \psi), \psi) \in \Gamma_q$. But from axiom *(3a)* we know that $\psi \wedge U(\neg S(\varphi, \psi), \psi) \rightarrow U(\neg S(\varphi, \psi) \wedge S(\varphi, \psi), \psi)$, so $U(\neg S(\varphi, \psi) \wedge S(\varphi, \psi), \psi) \in \Gamma_q$, which contradicts the consistency of Γ_q. The converse direction is symmetric.   □

By the above proposition, it follows that if a new point q' is added between points q and q'' at S or $\neg S$ stages that:

$$\{\varphi \mid \Gamma_q <_\varphi \Gamma_{q''}\} \cup \{\neg \psi \mid \Gamma_q <_\varphi \Gamma_{q''} \text{ and } \neg U(\psi, \varphi) \in \Gamma_q\} \cup$$
$$\{U(\gamma, \varphi) \mid \Gamma_q <_\varphi \Gamma_{q''} \text{ and } \varphi \in \Gamma_{q''}\} \subseteq \Gamma_{q'}$$

To finish the construction, let $\langle Q, < \rangle = \bigcup \{\langle Q_n, <_n \rangle \mid n \in \mathbb{N}\}$. Again it is not hard to check that $\langle Q, < \rangle$ is a countable dense unbounded linear order; the axioms $S(\top, \top)$ and $U(\top, \top)$ ensure unboundedness and $\neg U(\top, \bot)$ ensures density. So let us move straight way to showing that $\langle Q, < \rangle$ satisfies $\Delta \cup \{\neg \chi\}$.

Proposition 3.37. $\langle Q, <, v \rangle, q \models \gamma \Leftrightarrow \gamma \in \Gamma_q$.

> **Proof.**   By induction on the complexity of γ. We will only treat the case $\gamma = U(\varphi, \psi)$.
>
> [$\Rightarrow$] We will argue contrapositively. Assume $\neg U(\varphi, \psi) \in \Gamma_q$. If $\varphi \notin \Gamma_r$ for all $r > q$, then by the inductive hypothesis $r \not\models \varphi$ for all $r > q$. In this case, clearly $\langle Q, <, v \rangle, q \models \neg U(\varphi, \psi)$. So assume there is an $s > q$ such that $\varphi \in \Gamma_s$. Then there is some stage k such that $q, s \in Q_{k-1}$ and $\neg U(\varphi, \psi)$ is being treated. If at stage k there is no $q'', q < q'' < s$, such that $\neg \psi \in \Gamma_{q''}$, then a new point q' is added such that $q < q' < s$ and $\neg \psi \in \Gamma_{q'}$. By inductive hypothesis $s \models \varphi$ and $q' \models \neg \psi$, so $\langle Q, <, v \rangle, q \models \neg U(\varphi, \psi)$.
>
> [$\Leftarrow$] Assume $U(\varphi, \psi) \in \Gamma_q$. By construction, there is a stage j with $t \in Q_j, q < t$, such that

(i) $\varphi \in \Gamma_t$

(ii) $\forall r \forall s, q \leq r < s \leq t, \Gamma_r \prec_\psi \Gamma_s$

(iii) $\forall r, q < r < t, \psi \in \Gamma_r$

Furthermore, for any point v' added in the construction immediately between arbitrary points v and v''

$$\{\psi \mid \Gamma_v \prec_\psi \Gamma_{v''}\} \cup \{\neg\varphi \mid \Gamma_q \prec_\psi \Gamma_{v''} \text{ and } \neg U(\varphi, \psi) \in \Gamma_v\} \cup$$
$$\{U(\gamma, \psi) \mid \Gamma_v \prec_\psi \Gamma_{v''} \text{ and } \psi \in \Gamma_{v''}\} \subseteq \Gamma_{v'}$$

So it follows immediately from *(ii)* and *(iii)* that for all $q', q < q' < t$, $\psi \in \Gamma_{q'}$. By the inductive hypothesis, every $q' \models \psi$ and $t \models \varphi$, so $\langle Q, <, v \rangle, q \models U(\varphi, \psi)$. $\square$

It then follows that $\mathbf{Q_{SU}}$ is strongly complete with respect to $\langle \mathbb{Q}, < \rangle$.

4   Method Comparison

Throughout this paper we have made remarks contrasting the construction method to its more established model-theoretic counterpart. However, so far we have only sketched what the model-theoretic approach is, and we have not seen how the two approaches differ on any concrete examples. The purpose of this final chapter is to remedy this state of affairs. As a basis for method comparison, we will turn to the nicest application of the model-theoretic approach to topological completeness: **S4** on the Cantor Space. We will then see how the construction method can be applied to establish completeness, and discuss more generally the cases where the construction method has an advantage over the model-theoretic approach. Finally, we will speculate on further applications of the construction method in modal logics of space.

4.1   S4 on 2^ω, the Model-Theoretic Approach

As mentioned in the introduction, one current area of investigation in modal logics of space involves simplifying results found in McKinsey and Tarski's "The Algebra of Topology." This project seems to have started with Mints [11], where a new completeness proof of **S4** with respect to the Cantor Space is given. Since that time, a number of new proofs of completeness with respect to $\mathbb{R}$ and $\mathbb{Q}$ have been discovered, but none are quite as elegant as the Cantor Space proof. The reason for this is clear. **S4** has a large number of simple tree frame characterizations, and it is just more natural to transfer topological structure from one tree to another, than from a tree to some strict linear order. Let us examine how this model-theoretic transfer of topological structure from the Cantor Space to an **S4** tree works. As a note, the proof given below does not follow Mints [11], but rather a slightly different argument from Aiello, van Benthem, Bezhanishvili [3].

Fact 4.1.   **S4** *is complete with respect to the class of all rooted finite transitive and reflexive trees.*

Fact 4.2.   *Every* **S4** *frame* $\langle W, R \rangle$ *can be viewed as a topological space* $\langle W, \tau' \rangle$ *where* $\Delta_x = \{y \mid Rxy\}$ *and* $\mathcal{B} = \{\Delta_y \mid y \in W\}$ *serves as a basis for* τ'.

Recall that the domain of the Cantor Space $\langle 2^\omega, \tau \rangle$ is the set of all infinite sequences of 0s and 1s. Such infinite sequences $\sigma \in 2^\omega$ are called *branches* of the Cantor Space, and finite initial sequences $t \subset \sigma$ are called *nodes* of the Cantor Space. Let $B_t = \{\sigma \mid t \subset \sigma\}$. Then $\mathcal{B} = \{B_t \mid t$ is a node of the Cantor Space$\}$ serves as a basis for τ.

In outline, the model-theoretic completeness argument proceeds as follows:

- Assume $\nvdash_{\mathbf{S4}} \varphi$
- By **Fact 4.1**, φ can be refuted on a rooted finite transitive and reflexive tree $\langle W, R \rangle$
- By **Fact 4.2**, this tree can be viewed as a topological space $\langle W, \tau' \rangle$
- A continuous and open map is devised between $\langle 2^\omega, \tau \rangle$ and $\langle W, \tau' \rangle$
- This map is shown to be stronger than the suitable notion of modal equivalence
- So $\langle 2^\omega, \tau \rangle \nvDash \varphi$

Devising the continuous and open map between $\langle 2^\omega, \tau \rangle$ and $\langle W, \tau' \rangle$ is the step that requires work, and this is the nice part of model-theoretic proof. Essential use is made of the fact that $\langle W, R \rangle$ is a *finite* tree refuting φ.

Theorem 4.3. $\langle 2^\omega, \tau \rangle$ is complete with respect to **S4**.

Proof.   Assume $\nvdash_{\mathbf{S4}} \varphi$ and $\langle W, R \rangle$ is a rooted finite transitive and reflexive tree refuting φ. Intuitively, $\langle W, R \rangle$ can be viewed as tree in which some nodes are points in W and other nodes are sets of points $C \subseteq W$, where for every $x, y \in C$, Rxy and Ryx. We call such $C \subseteq W$ *clusters*. In other words, the nodes of the tree can be viewed as equivalence classes, some singletons and others proper.

Now let us view our Kripke frame $\langle W, R \rangle$ refuting φ as the topological space $\langle W, \tau' \rangle$. The first step is to devise a labeling between *nodes* of the Cantor Space and *points* in W. The labeling proceeds recursively as follows:

(1)   Label the root of the Cantor Space with a root of W (since the root of W may be a cluster, there could be more than one point to choose from).

(2)   If a node t of the Cantor Space is labeled with a point $w \in W$ then label every node $s \supseteq t$ of the branch $t000...$ with w.

(3)   If t is labeled with w, label the nodes occurring immediately to the right of the branch $t000...$ as follows: let $\{w_0, ..., w_k\}$ be an enumeration of Δ_w and label the node $t1$ with w_0, $t01$ with $w_1, ..., t(0)^k 1$ with w_k, $t(0)^k 01$ with $w_0...$, and so on.

Now we want to transform this labeling of nodes of the Cantor Space with points in W into a labeling of *branches* of the Cantor Space with points of W. We know that if a branch $\sigma \in 2^\omega$ is of the form $\sigma = t000...$ and t is labeled with w, then every node $s, t \subseteq s \subset \sigma$, is labeled w. In this case, we say w *stabilizes* σ. Otherwise, since all extensions $t \subseteq s \subset \sigma$ of a given node t labeled with w are labeled with members of Δ_w (and Δ_w is finite), we know that from some point on every node of σ is labeled with elements in some cluster $C \subseteq W$. With this in mind, let us define the following function $F : 2^\omega \to W$.

$$F(\sigma) = \begin{cases} w & \text{if } w \text{ stabilizes } \sigma \\ \rho(C) & \text{if } \sigma \text{ keeps cycling in } C \subseteq W, \text{ where } \rho(C) \text{ is an arbitrary member of } C \end{cases}$$

Lemma 4.4.   For any node t labeled with w , $F[B_t] = \Delta_w$.

As stated above, all extensions $s \supseteq t$ are labeled with points from Δ_w, so it follows immediately from the definition of F that $F[B_t] \subseteq \Delta_w$. For the converse, assume $w' \in \Delta_w$. Further assume $w' = w_j$ in the enumeration $\{w_0, ..., w_k\}$ of Δ_w. Then the branch $t(0)^j 1000...$ is labeled w'. So $\Delta_w \subseteq F[B_t]$. □

Lemma 4.5. F is a continuous and open map from $\langle 2^\omega, \tau \rangle$ to $\langle W, \tau' \rangle$.

Proof. F preserves opens: It suffices to check that F preserves basic open sets. However, by the above lemma F sends basic open sets B_t of $\langle 2^\omega, \tau \rangle$ to basic opens Δ_w of $\langle W, \tau' \rangle$.

F reflects opens: Let Δ_w be a basic open in $\langle W, \tau' \rangle$ and let $V = \bigcup \{B_t : t \text{ is labeled with a point } w \in \Delta_w\}$

Clearly V is open in $\langle 2^\omega, \tau \rangle$, so it suffices to check that $F^{-1}(\Delta_w) = V$. It follows easily from **Lemma 4.4** that $F(V) \subseteq \Delta_w$, so $V \subseteq F^{-1}(\Delta_w)$. Now assume $\sigma \in F^{-1}(\Delta_w)$. Then $F(\sigma) \in \Delta_w$ and we know by the labeling that from some point $t \subset \sigma$ on, all nodes $t \subseteq s \subset \sigma$ are labeled with members of Δ_w. So $\sigma \in B_t \subseteq V$ and $F^{-1}(\Delta_w) \subseteq V$.

Definition 4.6. Suppose two topological models $\langle X, \tau, v \rangle$ and $\langle X^*, \tau^*, v^* \rangle$ are given. A *topo − bisimulation* is a non-empty relation $T \subseteq X \times X^*$ such that if xTx^* then

- $x \in v(p)$ iff $x^* \in v'(p)$ for any propositional letter p
- if $x \in U \in \tau$ then $\exists U^* \in \tau^*$ such that $x^* \in U^*$ and $\forall y^* \in U^*$, $\exists y \in U$ such that yTy^*
- if $x^* \in U^* \in \tau'$ then $\exists U \in \tau$ such that $x \in U$ and $\forall y \in U$, $\exists y^* \in U^*$ such that yTy^*

Fact 4.7. *Let f be a continuous and open map between $\langle X, \tau \rangle$ and $\langle X^*, \tau^* \rangle$. Then given a valuation v^* on $\langle X^*, \tau^* \rangle$, the valuation $v(p) = f^{-1}(v^*(p))$ defines a total topo-bisimulation between the models $\langle X^*, \tau^*, v^* \rangle$ and $\langle X, \tau, v \rangle$.*

Fact 4.8. *If T is a topo-bisimulation between two models X, X^* such that xTx^*, then x and x^* satisfy the same modal formulas.*

To finish the proof, let v' be the valuation and w be the point in W such that $\langle W, \tau', v' \rangle, w \models \neg \varphi$. Then by **Lemma** 4.5 and **Fact** 4.7, we know that assigning $\langle 2^\omega, \tau \rangle$ the valuation $v(p) = F^{-1}(v'(p))$ defines a total topo-bisimulation between $\langle W, \tau', v' \rangle$ and $\langle 2^\omega, \tau, v \rangle$. So by **Fact 4.8**, $\langle 2^\omega, \tau, v \rangle, F^{-1}(w) \models \neg \varphi$. $\qquad \square$

4.2 S4 on 2^ω, the Construction Method

We have now seen how the model-theoretic approach can be used to show **S4** is complete with respect to the Cantor Space. Let us examine this proof from the perspective of the construction method.

Theorem 4.9. $\langle 2^\omega, \tau \rangle$ is complete with respect to **S4**.

Proof. Assume $\nvdash_{\mathbf{S4}} \chi$. Let $\mathcal{L}'$ be the language $\mathcal{L}$ restricted to subformulas of $\neg \chi$. Then there are only finitely many $\mathcal{L}'$ maximally consistent sets, each of which is finite. We will label nodes of the Cantor Space with $\mathcal{L}'$ maximally consistent sets as follows:

(1) Label the root r of the Cantor Space with Γ_r, an $\mathcal{L}'$ maximally consistent extension of $\{\neg \chi\}$.

(2) If a node t is labeled with Γ_t then label every node $s \supseteq t$ of the branch $t000...$ with Γ_t.

(3) If a node t is labeled with Γ_t then label the nodes occurring immediately to the right of the branch $t000...$ as follows. Let $\Sigma = \Diamond \psi_0, \Diamond \psi_1 ... \Diamond \psi_k$ be an enumeration of the diamond formulas in Γ_t. Label the node $t1$ with a $\mathcal{L}'$ maximally consistent extension of $\{\psi_0\} \cup \{\Box \varphi \mid \Box \varphi \in \Gamma_t\}$, the node $t01$ with

a maximally consistent extension of $\{\psi_1\} \cup \{\Box\varphi \mid \Box\varphi \in \Gamma_t\}$, $t(0)^k 1$ with a maximally consistent extension of $\{\psi_k\} \cup \{\Box\varphi \mid \Box\varphi \in \Gamma_t\}$, $t(0)^k 01$ with a maximally consistent extension of $\{\psi_0\} \cup \{\Box\varphi \mid \Box\varphi \in \Gamma_t\}$, ..., and so on.

Associate $\sigma \in 2^\omega$ with any $\mathcal{L}'$ maximally consistent set Γ which occurs infinitely many times on the nodes of σ (since there are only finitely many $\mathcal{L}'$ maximally consistent sets, such a Γ is guaranteed to exist). Assign $\langle 2^\omega, \tau \rangle$ the usual Henkin valuation. $\qquad\square$

Lemma 4.10. $\quad \langle 2^\omega, \tau, v \rangle, \sigma \models \varphi \Leftrightarrow \varphi \in \Gamma_\sigma.$

Proof. By induction on the complexity of φ. We will only treat the case $\varphi = \Box\psi$.

$[\Leftarrow]$ Assume $\Box\psi \in \Gamma_\sigma$. By examining the labeling procedure, it is clear that $s \subset t \Rightarrow \{\Box\psi \mid \Box\psi \in \Gamma_s\} \subseteq \{\Box\psi \mid \Box\psi \in \Gamma_t\}$. Since each $\mathcal{L}'$ maximally consistent set is finite, there must be some $t' \subset \sigma$ such that for all $t'' \supseteq t'$, $\{\Box\psi \mid \Box\psi \in \Gamma_{t'}\} = \{\Box\psi \mid \Box\psi \in \Gamma_{t''}\}$. It then follows from the labeling of branches that for every $\sigma' \in B_{t'}$, $\Box\psi \in \Gamma_{\sigma'}$. Furthermore, $\Box\psi \to \psi$ is an axiom (and ψ is in $\mathcal{L}'$ since it is a subformula of $\Box\psi$), so for all $\sigma' \in B_{t'}$, $\psi \in \Gamma_{\sigma'}$. Thus by the inductive hypothesis all $\sigma' \models \psi$. It follows that $\langle 2^\omega, \tau, v \rangle, \sigma \models \Box\psi$.

$[\Rightarrow]$ Assume $\Diamond\neg\psi \in \Gamma_\sigma$ and B_t is an arbitrary open such that $t \subset \sigma$. We know there is some $t' \supset t$ labeled with Γ_σ. Assume $\Diamond\neg\psi = \Diamond\varphi_j$ in the enumeration $\{\Diamond\varphi_0, \Diamond\varphi_1, ..., \Diamond\varphi_n\}$ of $\Diamond$ formulas in Γ_σ. Let $\sigma' = t'(0)^j 100....$ Then $\sigma' \in B_t$ and $\neg\psi \in \Gamma_{\sigma'}$, so by the inductive hypothesis, $\sigma' \models \neg\psi$. Thus $\langle 2^\omega, \tau, v \rangle, \sigma \models \Diamond\neg\psi$.

Let $\sigma_0 = 000....$ We know that σ_0 is associated with Γ_r and $\neg\chi \in \Gamma_r$, so it follows from the above lemma that $\langle 2^\omega, \tau, v \rangle, \sigma_0 \models \neg\chi$. $\qquad\square$

4.3 Method Comparison

When using the construction method to establish that **S4** is complete with respect to the Cantor Space, no mention is made of Kripke frame characterizations of **S4**, viewing these frames as topological spaces, establishing a homeomorphism between the Cantor Space and an arbitrary member of a class of Kripke frames, etc. The only relevant fact, that there are only finitely many maximally consistent sets, each of which is finite, is used in exactly the way you would expect given how $\Box$ and $\Diamond$ formulas are evaluated on branches of the Cantor Space. In this case, the construction method avoids a detour through Kripke semantics.

However, the mere streamlining of existing topological completeness proofs does not prove the utility of the construction method. This comes only when we moving to modal languages richer than $\mathcal{L}$, where the requisite Kripke frame characterizations are often very difficult to come by. In such cases, the construction method can be used to considerable advantage.

One instance where the construction method simplifies matters significantly is axiomatizing $\langle \mathbb{Q}, \tau \rangle$ in $\mathcal{L}^{\square FP}$. To get a sense of the difficulty of working model-theoretically, consider one of the common Kripke frame characterizations of **S4**. For example, recall that **S4** is complete with respect to the class of all finite transitive and reflexive trees. Call this frame class $\mathbb{F}$. To make the model-theoretic argument go forward, it is necessary to add structure to $\mathbb{F}$ to make the temporal axioms of $\mathbf{Q_{\square FP}}$ true. As it stands, axioms such as $F\varphi \rightarrow G(\varphi \vee P\varphi \vee F\varphi)$ (right-linearity) and $P\varphi \rightarrow H(\varphi \vee P\varphi \vee F\varphi)$ (left-linearity) need not be true on an arbitrary $\langle W, R \rangle \in \mathbb{F}$. The natural way to do this is to add an additional relation S to all $\langle W, R \rangle \in \mathbb{F}$ such that the temporal axioms (density, unboundedness, right linearity, left linearity, etc.) are true on S.

However, once this new frame class $\mathbb{F}'$ is conceived such that the axioms of **S4** and $\mathbf{Q_{FP}}$ are true on $\langle W, R, S \rangle \in \mathbb{F}'$, it still must be taken care that all the interaction principles between $\square$ and F and P are true on $\langle W, R, S \rangle \in \mathbb{F}'$. This is guaranteed to be difficult, since $\mathbf{Q_{\square FP}}$ is a somewhat expressive language and a tree containing two relations R and S bears little resemblance to the intended structure $\langle \mathbb{Q}, \tau \rangle$. Once this is done, it remains to be shown that $\langle \mathbb{Q}, \tau \rangle$ is modally equivalent to the relevant Kripke countermodel $\langle W, R, S \rangle \in \mathbb{F}'$.

In this case, rather than wrestling with Kripke frames it is better to go to the intended structure $\langle \mathbb{Q}, \tau \rangle$ directly. It is much simpler to force $\square$ and F and P to interact correctly on $\langle \mathbb{Q}, \tau \rangle$, as we did with the simple syntactic condition $<_{\square\psi}$, than establish that $\langle \mathbb{Q}, \tau \rangle$ is modally equivalent to an arbitrary member of some class of Kripke frames.

The difficulty in applying the construction method is not arriving at a structure with the right shape, but in ensuring that a harmony between the syntactic and the semantic has been established along the way. As we've seen, there must be a balance between the expressive power of the language and the richness of the topological structure being considered for the construction go smoothly. In cases such as $\mathcal{L}$ and $\mathcal{L}^{\square FP}$ on $\mathbb{R}$, where the language is not expressive enough to force the desired properties to hold on the eventual topological model, the construction appears to come to a halt.

Thus when considering when to use the construction method to establish spatial completeness results, there are two factors to consider. One is the richness of the desired structure, and the other is the expressive power of the language. In the case where the structure is rather simple, like $\mathbb{Q}$, the construction method can typically be applied to establish completeness. However, in these cases the model-theoretic method will usually work as well. It is the case when we move to richer languages that the construction method runs orthogonal to the model-theoretic approach. For while moving to a richer language almost always makes the model-theoretic approach more difficult (in terms of finding a suitable Kripke frame characterization, and transferring structure from the desired mathematical object), a stronger language often *simplifies* matters using the construction method. As was the case with $\mathcal{L}^{SU}$ on $\mathbb{Q}$, the strength of the language can actually make it *easier* to ensure that the model is being constructed correctly along the way, via a simple syntactic condition on maximally consistent sets. Thus when considering languages sufficiently stronger than $\mathcal{L}$, it seems reasonable to start with the construction method.

For instance, we believe the construction method might be useful for proving the completeness of a $\mathcal{L}^{SU}$ logic with respect to the real line. Furthermore, as we have seen in several cases, the construction method can be useful for establishing strong completeness results, while the standard model theoretic approach only yields completeness.

References

[1]   M. Aiello. Spatial reasoning: theory and practice, PhD thesis, University of Amsterdam, ILLC PhD series, DS-2002-02, 2002.

[2]   M. Aiello, J. van Benthem. A modal walk through space, Journal of Applied and Non-Classical Logics 12, 2002, no. 3/4, 319–363.

[3]   M. Aiello, J. van Benthem, G. Bezhanishvili. Reasoning about space: the modal way, Journal of Logic and Computation 13, 2003, No. 6, 889–920.

[4]   G. Bezhanishvili and M. Gehrke. A new proof of completeness of S4 with respect to the real line, ILLC prepublication series, PP-2002-06, 2002.

[5]   J. Burgess. Logic and time, Journal of Symbolic Logic 44, 1979, 556–582.

[6]   J. Burgess. Axioms for tense logic. I. "Since" and "until", Notre Dame Journal of Formal Logic, 23, 367–374, 1982.

[7]   D. de Jongh, F. Veltman. Intensional logics, 1986, lecture notes.

[8]   D. Gabbay, V. Shehtman, Products of modal logics I, Logic Journal of the IGPL 6, 1998, No. 1, 73–146.

[9]   H. Kamp. Tense logic and the theory of linear order, PhD thesis, UCLA, 1968.

[10]   J. McKinsey, A. Tarski. The algebra of topology, Annals of Mathematics 45, 1944, 141–191.

[11]   G. Mints. A completeness proof for propositional S4 in Cantor Space, in E. Orlowska, editor, Logic at Work: Essays dedicated to the memory of Helena Rasiowa, Physica-Verlag, Heidelberg, 1998.

[12]   G. Mints, T. Zhang. A proof of topological completeness of S4 on (0,1), Annals of Pure and Applied Logic 133, 2005, 231–245.

[13]   K. Segerberg. Modal logics with linear alternative relations, Theoria 36, 1970, 301–322.

[14]   V. Shehtman. A logic with progressive tenses, In M. de Rijke, editor, Diamonds and defaults: studies in pure and applied intensional logic, Kluwer Academic Publishers, Dordrecht 1993, 255–285.

Trio for Strings
223

Introduction

In the summer of 1958, between finishing his undergraduate degree at the University of California, Los Angeles, and starting graduate school at UC Berkeley, the twenty-two-year-old composer La Monte Young wrote *Trio for Strings* (1958–). Alternately viewed as a student provocation, a proto-Fluxus composition, and the first piece of minimalism, *Trio* is perhaps most accurately understood as a young person's working-through of modernist influences into a unique personal idiom.

As with any preliminary statement, *Trio* contains a number of curious features. Local notions of symmetry abound, but the overall form is asymmetric. The piece focuses on sustained harmonic groupings, yet contains long intervals of silence. Even silence itself is articulated, with tempo changes indicated on long periods of rest. While some of these curiosities can be "explained" through the structural lens of serialism, it is also evident that Young has drawn new conclusions from existing methods.

In this essay, we consider Young's novel use of twelve-tone methods in *Trio*, and suggest how his unique interpretation of these methods points toward his later signature style. Since Jeremy Grimshaw nicely details *Trio*'s twelve-tone structure in *Draw a Straight Line and Follow It*, his study of Young's life and work, we focus our efforts more on some ambient philosophical ideas surrounding minimalism arising out of modernist sources. As a particularly illuminating example, we consider the evolving performance history and the long development of a just intonation tuning for *Trio*, eventually leading to a canonical Dream Chord tuning in the early 2000s. This process sheds valuable light on Young's compositional approach taken as a whole.

We also compare Young's unique use of twelve-tone structures with some of Arnold Schoenberg's writing on twelve-tone methods from his Los Angeles years. In particular, we consider Schoenberg's notion of quantities preserved under transformation in twelve-tone music, and his related idea of a "unitary perception of musical space." We argue that *Trio* suggests interesting refinements of these notions, in the special case of identity under transformation. Young's novel interpretation points towards an alternative conception of twelve-tone music, with notions of continuity and constructability serving as primitive elements, in place of discrete orderings. These ideas, coupled with the audible structure of the harmonic series, provide a model for Young's development of minimalism and many of his subsequent compositional activities. To better trace these developments and appreciate the novelty of Young's conclusion, it is helpful to first consider the backdrop of European modernism in early 1950s Los Angeles.

At first glance, it seems surprising that one of the most novel interpretations of twelve-tone methods came from a recent college graduate in 1950s LA. However, at the age of twenty-two, in addition to being a working jazz musician performing with the likes of Ornette Coleman and Billy Higgins, Young was already near the forefront of modern music. This fact was appreciated by prominent figures of the time such as Leonard Stein and Milton Babbitt, who recognized the novelty of Young's early twelve-tone music and were supporters of his work. Indeed, over the next ten years Young would develop into one of the most important American composers, influencing John Cage, Terry Riley, Andy Warhol, Yoko Ono, the Velvet Underground, and a generation of artists to come.

While these prodigious efforts are no doubt a testament to Young's unique talents, when considering the evolution of his style it is also helpful to take into account a larger historical backdrop reflecting the impact of the Second World War and the development of Los Angeles as a cultural center. Schoenberg, who conceived the twelve-tone method in the 1920s, fled Nazi Germany in 1933 after having his position revoked at the Akademie der Künste in Berlin. Finding it difficult to secure a permanent position on the East Coast, in 1934 the sixty-year-old composer moved to Los Angeles with his young family, initially intending to teach composers in the growing film industry. At times to his chagrin, he spent most of his remaining years teaching: both privately, at the University of Southern California, and later at UCLA. While not generally perceived as an especially encouraging teacher, he took his role very seriously, especially in regard to educating students on the history of classical music.

Schoenberg's primary students during his Los Angeles years were Leonard Stein and Gerald Strang (although Cage attended classes at USC and UCLA, and Lou Harrison was a private student from 1941–42). While primarily known as a pianist, Stein studied composition with Schoenberg at both USC and UCLA, was his teaching assistant at UCLA from 1939 to 1942, and served as his personal assistant for the nine years leading up to Schoenberg's death in 1951. Though lesser known than Anton Webern or Alban Berg, Stein should be seen as one of Schoenberg's main students. They worked together closely for over a decade editing Schoenberg's later books and scores, in addition to Stein being a noted performer and promoter of his music in the area. Indeed, while Schoenberg's work initially faced resistance from prominent LA cultural institutions, late in life he enjoyed much greater public recognition, in no small part due to the efforts of his many students. By the 1950s, his ideas exerted a broad influence on the cultural landscape of the city. This is particularly evident in the case of Young.

In 1949, Young moved with his family from rural Utah to Los Angeles. In 1950, he began high school at John Marshall in Los Feliz. Known historically as something of a bohemian school, it has produced a wide range of performing artists over the years, from Terry Jennings to Leonardo DiCaprio. Both Young and the dancer Simone Forti have commented on the excellence of arts programs at Los Angeles high schools during this era (Forti was a student at Fairfax High when Young was at Marshall). Indeed, Clyde Sorenson, Young's primary music teacher at Marshall, studied with Schoenberg at UCLA, and Young has noted that his predisposition to twelve-tone methods likely stemmed from this early encounter. In addition to introducing young minds to modern music, it seems Sorenson also drew from Schoenberg's rigorous pedagogical style. Young once told me he took five semesters of harmony as a high school student alone.

Young enrolled at East Hollywood's Los Angeles City College in 1953. Although a two-year college, it was without question the most interesting music department in Los Angeles for an aspiring jazz musician and composer. In 1946, LACC became the first college in the nation to offer a degree in jazz, and subsequently it attracted many canonical figures as students, including Roy Ayers, Chet Baker, Eric Dolphy, and Charles Mingus. Known to have the nation's best jazz dance band (Young competed with Dolphy for second-chair saxophone), LACC also had very strong composition and piano performance programs, then headed by Leonard Stein. In fact, prior to working with Schoenberg, Stein himself had studied at LACC under the noted pianist Richard Buhlig. (It was Buhlig who encouraged both Cage and Stein to study with Schoenberg, and they attended his classes together at USC.)

Similar to his experience at Marshall High, Young's decision to attend LACC turned out to be highly fortuitous. Recognizing his potential, Stein took Young under his tutelage and became his most important early mentor and teacher. Many of Young's earliest surviving scores (*Scherzo* [1953], *Rondo* [1953], and *Wind Quintet* [1954]) were assignments for Stein's harmony and counterpoint classes. Furthermore, Young's first exploration of twelve-tone methods (*Five Small Pieces for String Quartet, On Remembering a Naiad* [1956]) was a product of his private compositional studies with Stein.

Stein's teaching broadly followed Schoenberg's approach, focusing on harmony and counterpoint without preference to style. Indeed, while Stein was a student of Schoenberg, it was Anton Webern who became Young's most important early compositional influence. Furthermore, as a prominent modern classical pianist, Stein was in a unique position to introduce Young to the larger world of musical modernism, including Stockhausen, Cage, Bartók, Nono, and others. Young recounts that Stein would pull records and scores from the library in advance of their meetings, and that he "literally introduced me to contemporary music."[1] Stein also played an important role in Young's professional development. Through his influence, Young was able to secure a scholarship to study at Stockhausen's 1959 class at the Darmstadt Festival.

As a result of his exemplary high school and community college experiences, Young was already close to the front of modern music by the time he reached UCLA in 1957. Indeed, when Milton Babbitt, the figurehead of American serialism in the 1950s and '60s, gave a guest lecture at UCLA in 1958, he was so thoroughly impressed by Young that he attempted to recruit him to Princeton in a series of letters.

This encounter with Babbitt is also notable in the context of *Trio*. Although the subject of Babbitt's UCLA talk is not clear, at the time he was developing the first group-theoretic analysis of twelve-tone methods and formal treatment of Webern's use of invariance. (It's interesting to note that Babbitt's younger brother, Alfred, was completing his PhD in algebra under Ellis Kolchin at Columbia during the same time.) As we shall see, invariance is a key tool employed in *Trio*, and indeed *Trio* is perhaps the most comprehensive use of the method in all of twelve-tone music.

While Young makes use of invariance in earlier pieces, his most systematic application roughly coincides with Babbitt's visit. In an August 1958 letter, Babbitt writes that he hopes Young can "come to Princeton, where I believe you would enjoy yourself. In the meantime, good luck with your string quartet. . . . Allow me to say what a great pleasure it was to see you and your music in Los Angeles."[2] It's worth considering whether Young and Babbitt discussed an early version of *Trio* during his UCLA visit, as there is no other string piece from this time.

Style and Idea

Schoenberg's last major work of writing was *Style and Idea*. Published in 1950, a year before his death, and edited by Stein, it collects essays written during his Los Angeles years and offers his final reflections on twelve-tone methods. No doubt inspired by his teaching activities, it has a pedagogical flair and likely bears a resemblance to presentations of classical and twelve-tone music given in his classes. Schoenberg noted that he tried to adjust his teaching style to suit more general audiences in California, where students often had less of a background. Indeed, "Composition with Twelve Tones," his final essay on twelve-tone methods, was modified from a lecture given at UCLA in March 1941.

Although Young never explicitly refers to *Style and Idea*, he was no doubt familiar with it, either directly or by way of Stein. Indeed, it is reasonable to expect that the presentation of twelve-tone methods Young received from Stein was influenced by the approach in *Style and Idea*. (Stein was Schoenberg's teaching assistant at the time of the "Composition with Twelve Tones" lecture and was heavily involved in all stages of the book's production.) As we shall see, one can trace some interesting parallels between Schoenberg's presentation of twelve-tone methods in *Style and Idea* and Young's unique application of these methods in *Trio*.

Distance in Equal Temperament

In conventional Western notation, the twelve notes in the chromatic scale function as primitive elements from which harmonic contexts are derived. The distance between sounds is understood in terms of the number of *steps* separating these notes. For instance, the distance between two adjacent keys on a piano is a half step; an octave is separated by twelve half steps, a fifth is separated by seven half steps, a fourth by five half steps, a major second by two half steps, and so on. Steps provide discrete, independent units of measurement between sounds.

Consequently, the distance between musical intervals is understood through the addition and subtraction of units. The distance spanned by ascending a fifth (seven half steps) and then a fourth (five half steps) is $7 + 5 = 12$ half steps, or an octave, while the difference between a fifth and a fourth is $7 - 5 = 2$ half steps, or a major second. While the meaning(s) of notes or chords depends on the harmonic contexts in which they appear, these contexts themselves are formed of atomistic pieces: discrete and "equally spaced" notes with ordinal position.

Serialism

Serialism can be viewed as a method of successive variation based on orderings of these atomic elements. The starting point of serial music is a given ordering of the twelve notes in the chromatic scale, called the *tone row T*. Basic compositional materials are drawn from a specified collection Ω of permutations of T, including:

(i) The *transposition T_n* of T. This permutation shifts T up n half steps. Assuming octave equivalence, we may restrict to $0 \leq n \leq 11$[3].

(ii) Writing T backwards. This is called the *retrograde R* of T.

(iii) Let $(a_1, a_2, \ldots, a_{12})$ list the order of notes in T. For $1 \leq i \leq 11$, let γ_i be the number of half steps separating a_i and a_{i+1} (where $\gamma_i > 0$ if a_{i+1} is a note above a_i, and $\gamma_i < 0$ otherwise). Let $\beta_1 = a_1$, and β_{i+1} be the note $-\gamma_i$ half steps from β_i. The permutation $(\beta_1, \beta_2, \ldots, \beta_{12})$ of T is called the *inversion I* of T.

Certain compositions of permutations are also included in Ω, for instance $RI_9 = R \circ I \circ T_9$, the retrograde inversion of T up nine half steps. This permutation is formed by first transposing T up nine half steps (T_9), then inverting the result (I_9), then finally taking the retrograde of I_9 (RI_9).

Slightly generalizing this process, the collection of permutations Ω forms a mathematical object called a *group*. From this perspective, basic compositional materials are drawn from a permutation group acting on the notes in the chromatic scale.

Symmetry in Serialism

Given the above presentation, it is not immediately clear why this particular collection Ω of permutations is favored for compositional materials over the (many) other possibilities. When considering this question, it should be noted that Schoenberg did not conceive of serialism in terms of permutations of a set, but rather as symmetries of some more structured kind of "musical space."[4]

Schoenberg notes that, in twelve-tone music, "the basic set [the tone row] is used in diverse mirror forms."[5] He likens these mirror forms to symmetries of two- and three-dimensional Euclidean space, such as reflections and rotations, which are fundamental to both visual art and our basic perceptual faculties. Given the primordial nature of these symmetries, Schoenberg argues that we view them as a part of our discernment of the world: "these musical ideas correspond to the laws of human logic; they are a part of what man can apperceive."[6] In compositional terms, he sees a long precedent in the use of these forms, though largely from composers preceding the Romantic era: "the composers of the last century had not employed such mirror forms as much as the masters of contrapunctal times; at least, they seldom did so consciously."[7]

Schoenberg conceives of these mirror forms as symmetries of a two-or-more-dimensional "musical space" where "elements of a musical idea are partly incorporated in the horizontal plane as successive sounds, and partly in the vertical plane as simultaneous sounds."[8] From the assumption that these basic symmetries are immediately given, he concludes in twelve-tone music:

> THE TWO-OR-MORE-DIMENSIONAL SPACE IN WHICH
> MUSICAL IDEAS ARE PRESENTED IS A UNIT. Though the
> elements of these ideas appear separate and independent to the
> eye and the ear, they reveal their true meaning only through their
> cooperation, even as no single word alone can express a thought
> without relation to other words. All that happens at any point of this
> musical space has more than a local effect. It functions not only in

its own plane, but also in all other directions and planes, and is not without influence even at remote points.[9]

In this way, the obvious horizontal and vertical dimensions of music are in fact projections of a single, higher-dimensional and unified musical space upon which the mirror forms act. Schoenberg stresses that mirror forms act on all dimensions of this space simultaneously and are "not without influence even at remote points." He frames this in part as a distinction between twelve-tone music and its tonal counterparts.

Of course, it is highly unlikely that Schoenberg conceived "musical space" as some sort of abstract entity or independent object of study. (At the very least, he never attempted to work out such an idea.) Rather, it was a convenient expository tool for describing how mirror forms operate in twelve-tone contexts. Indeed, it was only late in life that Schoenberg was sometimes put forward as a "mathematical composer," much to his own confusion. His wife Gertrude told Babbitt that Schoenberg was baffled by this assertion: "My husband didn't know any mathematics, and didn't even know of what they were accusing him."[10]

Symmetry in Tonal Music

In *Style and Idea*, Schoenberg describes tonality as a local concept, in the sense that everything occurring in a piece implicitly refers back to its tonal center. This creates a sort of nonhomogeneous space of relations, with certain harmonic configurations appearing much larger in the foreground and others receding to a distant horizon. Schoenberg suggests that by removing the tonal notion of a root, twelve-tone methods can offer a more global, isotropic and "unitary" perspective. In this setting, mirror forms serve as true symmetries of a global ambient "space in which musical ideas are presented." The symmetries have "more than a local effect. [They] function not only in [their] own plane, but also in all other directions and planes, and [are] not without influence even at remote points."

By contrast, these same forms may not serve the same purpose in tonal settings, where their function is limited by the tonal contexts in which they appear. Either the relevant notion of symmetry only applies to a single musical dimension, or its effects are limited by the nonhomogeneous and local nature of the ambient environment.

Schoenberg additionally makes the argument that in twelve-tone contexts mirror forms preserve quantities, both musically and perceptually. He writes:

> *The unity of musical space demands an absolute and unitary perception.* In this space, as in Swedenborg's heaven (described

in Balzac's *Seraphita*) there is no absolute down, no right or left, forward or backward. Every musical configuration, every movement of tones has to be comprehended primarily as a mutual relation of sounds, of oscillatory vibrations, appearing at different places and times. To the imaginative and creative faculty, relations in the material sphere are as independent from directions or planes as material objects are, in their sphere, to our perceptive faculties. Just as our mind always recognizes, for instance, a knife, a bottle or a watch, regardless of its position, and can reproduce it in the imagination in every possible position, even so a musical creator's mind can operate subconsciously with a row of tones, regardless of their direction, regardless of the way in which a mirror might show the mutual relations, which remain a given quantity.[11]

Although music proceeds sequentially in time, Schoenberg argues that a row is a given quantity, which can be intuited in abstraction of temporal categories, and regardless of how mirror forms are applied. A "unitary perception" precedes specific position or orientation, in both the material and musical spheres. For the ideal listener, this quantity may be experienced "regardless of direction, regardless of the way in which a mirror might show mutual relations."

Hence, while serialism may naturally be viewed as a method of variation, in his later years Schoenberg also highlights quantities preserved under transformation as a distinguishing feature, both musically and perceptually. This is intriguing in the context of Young, where the twin ideas of preserving quantities and the experience of identity under transformation provide a key backdrop to his interpretation of twelve-tone methods and his later development of minimalism.

Invariance

The conception of twelve-tone methods preserving quantities is perhaps most clearly observed in the compositional tool of invariance. While serial methods are naturally conceived in terms of orderings and hence the sequential in time, Schoenberg notes that there are other perspectives: "a basic set of twelve tones can be used in either dimension, as a whole or in parts."[12] It is possible to apply mirror forms to vertical dimensions and study harmonic quantities preserved under these operations. It was Schoenberg's student Anton Webern who most clearly articulated this vertical dimension of twelve-tone music.

In his application of twelve-tone methods, Webern often subdivided the tone row into a collection of pitch sets (for instance, into four sets containing three successive notes in the row, or two sets of three notes and three of two notes, etc.). This collection of pitch sets provides a harmonic overlay to the ordinal presentation of the row. Webern then selected mirror forms to preserve harmonic quantities as well as sequential features. One key property Webern sought was *invariance*: given a grouping of the row into pitch sets, a given set is invariant under a serial operation if it is preserved by the operation—in other words, if the grouping appears in the row after the serial operation has been applied. Using Schoenberg's conception of musical space, one might see an invariant set as a harmonic "subspace" preserved by the operation.

As the grouping of the row into pitch sets concerns solely harmonic quantities, which are by their nature simultaneous, the sequence of the notes in the pitch set is immaterial, as is the location of the pitch set in the initial ordering. This ordering structure is not observable in the vertical dimension, just as the harmonic groupings themselves are not evident when projecting onto an ordinal dimension.

Schoenberg argues that the strength of twelve-tone methods stems from the context it supplies, and the power of mirror forms applied in all dimensions. Unconstrained by tonal contexts, these forms are free to operate on all dimensions simultaneously: "in all directions and planes" and with "influence even at remote points" while preserving musical ideas or expressions.

Young and Webern

Young notes that "while Schoenberg may have discovered stasis deep from within his subconscious, it was Webern who focused on and developed this phenomenon."[13] Or, in a slightly more pointed formulation from the late 1960s: "Schoenberg used row technique in a far more naive way—it was not as strictly, in an audible way, related to the musical result, as it is in the work of Webern."[14] Webern's attempts to manifest formal structure in audible terms was one of Young's key compositional influences. Indeed, while Young's formative teachers studied with Schoenberg, it was Webern's spare palette, static surfaces, and elegant use of formal structure that most captured the youthful composer's imagination.

Young's first twelve-tone composition, *Five Small Pieces for String Quartet, On Remembering a Naiad*, was inspired by Webern's set of *Six Bagatelles* for string quartet (1913) and his *Five Pieces for Orchestra* (1911–13). In *For Brass* (1957) and *For Guitar* (1958), Young continues Webern's exploration of stasis through twelve-tone methods, including the use of invariance. *Trio for Strings* can be seen as the culmination of this approach, possibly exhausting this avenue of exploration for Young. Shortly after this piece, he abandons the use of twelve-tone structures, choosing to instead explore stasis through new methods of his own invention while consistently returning to the same harmonic material investigated in his early pieces.

It should be noted that while Young's early work is highly indebted to Webern in structural terms, *Trio* sounds little like Webern (or any other twelve-tone music). While Webern's music is characterized by brevity and concision (perhaps so its elegant formal structures are better perceived), Young stretches these same formal methods far beyond any hopes of explicit recognition. Instead, the methods are indicative of a qualitative experience—of stasis as a subjective personal experience.

In this way, one might view Young's use of long sustained tones in a twelve-tone context as a notable extension of Webern's investigation of stasis and indeed as an extension of twelve-tone methods more generally. Young replaces the horizontal dimension in Schoenberg's musical space with a plane of subjective experience upon which mirror forms act. This wedding of the subjective experience of duration to formal methods is likely the key conclusion that allows Young to move beyond twelve-tone methods after *Trio* and to investigate stasis in his own terms. Indeed, this model of wedding subjective experience to formal methods, already present in *Trio*, is a key component of Young's development of minimalism in the 1960s.

Invariance in Trio

Although a strict twelve-tone composition, *Trio*'s primary structural unit is not the tone row, but rather its subdivision into pitch sets. While this is clear from an analysis of the score, it is manifestly obvious as an auditory experience. The piece is made up of sustained harmonic groupings and silences, with only the implicit melodic content of one note following another at various times. In the first performed version, the exposition of the row itself takes over twelve minutes to complete (and closer to forty minutes in preliminary sketches). This removes customary notions of melody and, at least locally, reduces identity to pure, simultaneous harmonic quantities. In fact, this notion of identity is explored through a *single harmonic grouping*, later referred to as a Dream Chord.

In *Trio*, the tone row is divided into the harmonic constellations:

$$\{\{C\sharp, E\flat, D\}, \{B, F\sharp, F, E\}, \{B\flat, A\flat, A\}, \{G, C\}\} \qquad (*)$$

For simplicity, at times we will express these constellations in prime form as: $\{\{C\sharp, D, E\flat\}, \{B, E, F, F\sharp\}, \{A\flat, A, B\flat\}, \{C, G\}\}$. The twelve-tone structure restricts the number of possible harmonic configurations to intervals occurring in the tone row and their inversions. Invariance further restricts this already limited harmonic palette. Indeed, one of Young's primary compositional strategies in *Trio* is the use of twelve-tone methods to ensure as *few harmonic groupings as possible occur*.

The overall twelve-tone structure of *Trio* is:

$$T \to I_9 \to RI_9 \to I_4 \to RI_4 \to T \to Coda.$$

Given Young's subdivision of the row (∗), **Table 1** below specifies the harmonic collections invariant under the available mirror forms.

Table 1: Harmonic constellations invariant under mirror forms

	{C♯, E♭, D}	{B, F♯, F, E}	{B♭, A♭, A}	{G, C}
I_1/RI_1	×	×	×	×
I_2/RI_2	✓	×	×	×
I_3/RI_3	×	×	✓	×
I_4/RI_4	×	×	✓	×
I_5/RI_5	×	×	×	✓
I_6/RI_6	×	×	×	×
I_7/RI_7	×	×	×	×
I_8/RI_8	×	✓	×	×
I_9/RI_9	✓	×	✓	×
I_{10}/RI_{10}	×	×	×	×
I_{11}/RI_{11}	×	×	×	×
I_0/RI_0	×	×	✓	×

Analyzing this information, Young's compositional approach is clear. The *only* row operations preserving two harmonic groupings ({C♯, E♭, D}, {B♭, A♭, A}) from (∗) are I_9 and RI_9. Similarly, I_4 and RI_4 preserve a grouping ({B♭, A♭, A}) common to I_9 and RI_9. Hence given the initial choice of harmonic groupings (∗), which clearly dictated Young's construction of the row, *Trio* has as little harmonic material as possible for all twelve-tone compositions containing at least two distinct (non-identity) mirror forms.

Given Young's strategy of restricting harmonic groupings through formal methods, it's worth noting that a slightly different choice of a row with added Dream Chord structure would exhibit even less variation:

$$\{\{B, C♯\}, \{C, F, F♯, G\}, \{E♭, A♭, A, B♭\}, \{D, E\}\} \qquad (∗∗)$$

Given this presentation, the mirror forms I_5/RI_5 leave *every* set invariant, and hence fix the entire harmonic space. Nonetheless, in general terms, it is interesting that Young employs a method of variation to ensure as little variation as possible occurs. This draws focus on the idea of identity under transformation, a central concept we shall explore for the rest of the essay.

Dream Chords

The intervals present in the set $\{B, E, F, F\sharp\}$ (the second harmonic collection in (*) above) play a central role in Young's entire musical output. This collection can be used to form what he calls a Dream Chord: a fifth ($\{B, F\sharp\}$), containing a nested fourth ($\{B, E\}$), and a semitone in between ($\{E, F, F\sharp\}$). Indeed, each of the four pitch sets in (*) are subsets of a Dream Chord, and the row itself is devised to create a successive filling in and thinning out of this Dream Chord structure (one could also interpret the lengthy silences occurring in the piece as subsets of a Dream Chord). Perhaps Young favored this filling-in and thinning-out structure over the more symmetric option (**) presented above.

Dream Chords were generative for Young even before he became a composer. In an interview with Richard Kostelanetz from the late 1960s, Young refers to the sound of humming electrical wires outside his childhood log cabin in Idaho as an important formative musical experience: "There is the 'Dream Chord,' which I used to hear in the telephone poles, which is the basis for the *Trio for Strings*. It is, for instance, *G, C, C♯, D*."[15]

Even after abandoning twelve-tone structures, Young continues to write pieces made up of the same Dream Chord material as *Trio*. For instance, *The Four Dreams of China* (1962) and *The Subsequent Dreams of China* (1984) are composed entirely of the four notes in a Dream Chord, heard in different voicings. Two of the *Four Dreams* reference the early childhood inspiration of humming power wires. For instance, one is titled *The Second Dream of the High Tension Line Stepdown Transformer*.

Unlike *Trio*, *Four Dreams* has no fixed score determining abstract architectural properties in advance. Rather, as is typical of much of Young's later work, the pieces are given generatively through a set of rules specifying how notes may enter and exit. In the 1980s, new *Melodic Versions* of the *Four Dreams* were developed, likely inspired by Young's study of raga music, with a set of "melodic rules" determining the entrances and exits of notes in the Dream Chord. Any individual performance develops in time according to the performer's continued attention, and through the configurations occurring up to present. One can naturally view the *Dreams* as casting light on the harmonic framework of *Trio* and its use of stasis, but through new methods better suited Young's compositional goals. In fact, we will also see the reverse process holding, with the performance history of *Four Dreams* providing a key backdrop to the development of a just intonation tuning for *Trio*.

While the Dream Chord serves as *Trio*'s basic structural unit, and invariance is a key formal tool, one should also note that many different compositional approaches coexist in the piece, sometimes in idiosyncratic ways. While *Trio* is a strict twelve-tone composition, it also has sonata form, and Young cites Indian and Japanese musical influences. This leads to some interesting tensions in the piece. Young goes to great lengths to ensure many local notions of symmetry hold, both in the ordinal and harmonic dimensions, yet the classical convention of the coda obstructs any global notion of symmetry. In fact, in spite of its rigorous twelve-tone structure, Young describes *Trio* as

> a rather tonal piece. It's in some sort of C . . . probably . . .
> C-minor . . . . It doesn't start there, but it gets there: in the cadence
> of the exposition and in the cadence of the recapitulation and in the
> cadence of the coda.[16]

He suggests that the long fifth *{C,G}* sustained at the end of the exposition, which serves as the first real consonance of the piece, reveals an implicit key.

Similarly, while Young highlights the "emphasis on harmony to the exclusion of any semblance of what had generally been known as melody"[17] as one of *Trio*'s distinguishing features, some of his later notions of melody are already present in the piece.

He writes:

> In *Trio for Strings*, there is no melody as each tone is separated by
> silence from its preceding and succeeding tones in the same voice.
> The texture is contrapuntal in that long, sustained tones overlap
> in time. Melody exists only in the sense that one remembers and
> identifies events that have taken place over long periods of time.[18]

On the other hand, the order in which the notes enter and exit is carefully determined and frequently match the later "melodic rules" in *The Four Dreams of China*. For instance, the note serving as the semitone in between the major second often enters as a disturbance of simpler intervals, and the *F* in the second constellation initially appears as an artificial harmonic in a brief pulse. Furthermore, the Dream Chords are sequentially transposed in the initial presentation of the row. If the four constellations were to be fully filled out in their Dream Chord structure, it would lead to a progression of thirds (or possibly one sixth). This is somewhat surprising, since Young highlights the absence of thirds as a characteristic feature of *Trio* and much of his early work. Again, while this doesn't reveal any explicit melodic content, it does elucidate some latent sense of progression present when listening to the piece.

Symmetry in Trio

In his later years, Schoenberg highlighted quantities preserved under transformations as a key feature of twelve-tone music, both in compositional and perceptual terms. One can see Young's application of twelve-tone methods as clarifying this idea, in the special case of identity under transformation. As discussed above, in *Trio*, twelve-tone methods are employed to explore a harmonic quantity, the Dream Chord, in all of its mirror forms. Unlike Schoenberg's conception of twelve-tone operations acting as symmetries on an underlying multidimensional space, *Trio* at least initially appears to reflect a single dimension of simultaneous harmonic quantities, further restricted to symmetries of a specific harmonic subspace.

At first this seems a little curious. One might ask: what is the intent of employing elegant formal methods to ensure as little as possible occurs? Also, if the result of applying mirror forms are the same harmonic configurations you started with, how is one supposed to perceive this process of transformation? Or, stated slightly differently: What is the intent of applying invariance to the exact object of study, the Dream Chord, as opposed to some slightly more ambient environment it sits in (like Schoenberg's "musical space")? I asked this question to Young once and he replied: "Well, I think you can hear it."[19] As a younger composer, I thought about this statement for a long time, and it was the backdrop for pieces of my own, such as *flag (iv)* (2007), which combine symmetries of a restricted harmonic palette with some spare melodic aspects.

For Schoenberg, unitary perception concerns the intuition of the row as a given quantity, regardless of orientation. In Young's work, the use of extended duration stretches the row beyond any recognizable figure and indeed beyond the conventional horizontal plane of twelve-tone music. In its place, Young adds a subjective dimension of duration and identification. These twin dimensions of harmonic quantities and perceptual identification constitute Young's version of musical space, upon which the mirror forms act. Note this is a reversal of the typical presentation of twelve-tone music—here, notions of continuity and constructability serve as primitive elements rather than discrete orderings of elements. This project of wedding formalism to subjective experience is essential to Young's work and becomes more explicit with the introduction of just intonation.

Just Intonation

At an atomistic level, sounds in equal temperament are represented sequentially by notes in a scale. Another way of viewing sounds is in terms of their respective frequency ratios. For instance, the three frequencies $800Hz$, $600Hz$, and $400Hz$ (800, 600, and 400 cycles per second) can be expressed by the ratio 800:600:400 or equivalently by the ratio 4:3:2. Just intonation is a system of tuning where sounds are related to each other in terms of such simple whole-number ratios.

In just intonation a single ratio may be realized in infinitely many ways. For instance, 4:3:2 can be realized by the frequencies 400*Hz*, 300*Hz*, 200*Hz*; 800*Hz*, 600*Hz*, 400*Hz*; or even 0.8*Hz*, 0.6*Hz*, 0.2*Hz*. Furthermore, assigning a single frequency suffices to determine pitch values for the entire group.

In this system, the primordial elements are not individual discrete notes, but rather *simultaneously given* frequency ratios. Unlike conventional notation, with its emphasis on independent units of measurement, in just intonation sounds attain a value only in relation to other frequencies occurring simultaneously. (Of course, one might say that an individual frequency serves as an identity, but this would be represented by the ratio 1:1 rather than any specific frequency.) In this way, just intonation represents an indecomposable continuity of experience, rather than adding a harmonic overlay on top of irreducible ordinal units. The representation of harmonic quantities as primitive elements closely accords with Young's approach in *Trio*, although he was not familiar with this system at the time.

In just intonation, distance is understood directly in terms of *ratios of frequencies*, and hence is given multiplicatively (or geometrically), rather than additively. An octave is defined as two frequencies whose ratio is 2:1, a fifth as two frequencies whose ratio is 3:2, a fourth as two frequencies whose ratio is 4:3, and a major second as two frequencies whose ratio is 9:8. To find the distance spanned by ascending two consecutive intervals it suffices to multiply the two ratios, and to find the distance separating intervals it suffices to divide the larger interval by the smaller one. So, for instance, the distance traversed by first ascending a fifth (3:2) and then a fourth (4:3) is the interval $\frac{3}{2} \times \frac{4}{3} = \frac{2}{1}$, or an octave. Similarly, the difference between a fifth and a fourth is $\frac{3}{2} \div \frac{4}{3} = \frac{9}{8}$, or a major second.

Just Intonation and Equal Temperament

There are some obvious differences between just intonation and equal tempered systems. For instance, in equal temperament ascending twelve fifths is equivalent to ascending $12 \times 7 = 7 \times 12 = 84$ half steps, or seven octaves. In just intonation, this amounts to saying that $(3/2)^{12}$ (twelve fifths) = $(2/1)^7$ (seven octaves), a false numerical statement. This discrepancy, referred to as the Pythagorean Comma, demonstrates a difficulty about the division of octaves known since antiquity.

Nonetheless, it is natural to wonder whether any apparent differences between these systems could be reconciled by simply assigning some geometric distance to the independent unit of measurement in equal temperament, the half step. While a natural idea, this does not end up working. If one were to choose a geometric distance to assign a half step, the most reasonable choice would come from dividing an octave into twelve equal parts. By the above reasoning, this would mean assigning a half step the geometric distance of $^{12}\sqrt{2}$.

This leads to some unexpected consequences. First of all, it is a classic result that there are no whole numbers m, n, and $p > 1$ such that $(m/n)^p = 2$. This implies that just intonation and equal temperament will disagree on every interval besides the octave, and even more that no equal-tempered interval besides an octave is a just intonation interval (since $^p\sqrt{2}$ is not a fraction). Second, according to Young's tuning theory (discussed below), it is not possible to tune to an irrational interval like $\sqrt{2} : 1$ due to the two frequencies having completely distinct overtone series.

Of course, in practice, tuning a piano in equal temperament does not attempt to produce equally spaced intervals for each half step. Rather subtle adjustments are made so the most consonant intervals are closer to being "pure" or "just" intervals. Nonetheless, this is an interesting case where theoretical or philosophical issues are manifested in actual auditory experience.

The Overtone Series, Harmony and Just Intonation

While we have now spent considerable time discussing the relationships between sounds, in theory every *individual note* performed on a piano is composed of infinitely many sounds and contains all intervals occurring simultaneously. The lowest of these sounds, the *fundamental*, is "performed," with additional frequencies, or overtones, sounding in a regular series above the fundamental. In fact, the relative presence or absence of these overtones determines the unique timbre of an instrument; this in turn explains why two instruments performing the same pitch each sound unique.

It is simplest to conceptualize this process by considering stringed instruments. The length of the entire string arching back and forth sounds the fundamental frequency. The overtone series corresponds to regular subdivisions of the vibrating string into successively shorter standing waves. In the present case, either the entire string vibrates back and forth, or the string may divide into two equal parts with a fixed (or stationary) node at the midpoint, or into three parts with two fixed nodes at one-third and two-thirds the length of the string, or into four parts, and so on.

The process of subdivision described above creates either one standing wave of the entire length of the string (the fundamental), or two standing waves of half the length of the string (the first overtone), or three standing waves of one-third the length of the string (the second overtone), or four standing waves, and so forth. Although a fixed string may vibrate in many different patterns, each of these patterns must occur as one of the regular subdivisions described above. The greater the number of parts the string is divided into, the shorter the standing waves, and the higher the corresponding overtone sounding above the fundamental pitch. In theory, this process of subdivision is infinitely extendable, and a string fixed at both ends set in motion may vibrate in any one of these patterns. In an ideal sense, an individual note from a stringed instrument consists of infinitely many of these related frequencies occurring simultaneously, of which the performed pitch is the lowest and typically most prominent.

The various subdivisions of a string into standing waves corresponds to a regularly occurring series of frequencies sounding above the fundamental pitch. Given a fundamental frequency of a cycles per second, the n-th overtone has frequency $a(n + 1)$ cycles per second. (Note the $n+1$-th overtone is also referred to as the n-th partial, so the n-th partial has frequency an cycles per second.) For instance, if the fundamental frequency is 200 Hz, then the first overtone (and second partial) is 400 Hz, the second overtone is 600 Hz, the third overtone is 800 Hz, and so on.

Notice that the frequency ratios of the first few overtones are familiar from our discussion of just intonation intervals. The frequency ratio of the first overtone and fundamental is 2:1, or an octave; the frequency ratio of the second overtone and first overtone is 3:2, or a "just" or "pure" fifth; and the frequency ratio of the third overtone and the second overtone is 4:3, or a just fourth. In general, the frequency ratio between the n-th and $n-1$-th overtone is $n + 1$:n.

Many theories of harmony are derived from the overtone series. In a standard presentation, the most consonant intervals (an octave, a fifth, a fourth) correspond to the frequency ratios of the first few successive overtones (2:1, 3:2, 4:3). This is explained by the fact that the overtone series of these respective frequencies are closely related. From the above discussion, if an interval is represented by the frequency ratio x:y, then xyn will be a common partial of both frequencies, for every positive integer n. So, two frequencies forming an octave share every second overtone of the higher frequency in common, while frequencies in a pure fifth share every third overtone of the higher frequency in common, and so on.

Conversely, intervals whose common overtones are widely separated, or nonexistent, are typically viewed as dissonant. For instance, in equal temperament the interval corresponding to six half-steps is called a *tritone*. This interval purports to divide an octave into two equal parts, which corresponds to a frequency ratio of $\sqrt{2}$:1. As noted above, $\sqrt{2}$ cannot be written as a fraction, and as a consequence the series set up by these two frequencies will share *no common overtones*. (In fact, Young has claimed it is impossible to tune such an interval, since there is no periodicity in the composite waveform.) This is commonly offered as an explanation for why the tritone is perceived as dissonant.

From this discussion, another way of viewing just intonation immediately arises: as a system of harmonic relations derived from the overtone series. Since just intonation intervals are whole number ratios of frequencies, *every* just intonation interval can be seen as occurring simultaneously in the overtone series of a single note. In this way the entire harmonic spectrum, and by extension every interval, can be viewed as existing in each moment, it is just a matter of perceiving it. This idea becomes key to Young's compositional process.

Dream Music and Vertical Hearing

Young's disposition towards continuity and harmonic quantities as primitive elements is closely mirrored in the theory of just intonation. Indeed, upon discovering this system in the early 1960s, the audible structure of the harmonic series became a central principle of organization in Young's music. Beginning with *The Tortoise, His Dreams and Journeys* in 1964, all of Young's work is written in just intonation. The initial performances of *Tortoise* involved voice and string instruments sustaining just intonation intervals over a continuous fundamental provided by an oscillator. A performance group referred to as the Theatre of Eternal Music was assembled for the rehearsal and presentation of this piece and related "Dream Music." In a 1964 concert program Young writes:

> In Dream Music there is a radical departure from European and even much Eastern music in that the basis of musical relationship is entirely harmony. Not European harmony as textbooks have outlined it, but the intervallic proportions and acoustical consequences of the particular ratios which sound concomitantly in the overtone series when any simple fundamental is produced. Melody does not exist at all (The Disappearance of Melody) unless one is forced to hear the movement from group to group of various simultaneously sounded frequencies derived from the overtone series as melodic because of previous musical conditioning. Even before the first man moved successively from one frequency to another (melody if you like) a pattern for this movement, that is the relationship of the second frequency to the first was already predetermined (harmonically) by the overtone structure of the fundamental of the first sound.[20]

In this conception, all compositional materials are both immediately given and continuously present as overtones of the fundamental, and are constructed explicitly through ongoing auditory detection. Young refers to this continuing process of locating regions of the overtone series as *"Vertical Hearing (or Hearing in the Present Tense)."*[21] It is one of his key theoretical constructs, along with the closely related notion of tuning as a function of time, which says that the degree of precision of an interval corresponds to the number of observed cycles of its composite waveform. In both cases, intervals, and by extension compositions, are given as developing in time through an ideal performer's continued attention.

This process of composing and performing through the audible structure of the harmonic series can be seen as casting light on changing subsections of a continuously present harmonic spectrum. Crucially, in this conception intervals are treated as synthetic constructions occurring in time through audible detection, rather than *a priori* entities. As in *Trio*, this method of composition pairs primitive notions of continuity and constructability. Unlike conventional twelve-tone music, the formal system is now explicitly tied to the performer's continued subjective awareness. In some respects, this new compositional framework bears a likeness to an aesthetic version of Nietzsche's parable of the eternal recurrence, where all things that can or will occur are present in a given moment, it is just a matter of having the aesthetic disposition to perceive it.

Pre-Tortoise Tunings

As early as 1965, Young considered the possibility of placing the Dream Chord in just intonation, which he felt would more faithfully reproduce the sound of the step-down transformers from his youth. Most intervals appearing in a Dream Chord have a canonical representation in terms of the harmonic series; a fifth is represented by the ratio 3:2, a fourth by 4:3, and a major second by 9:8. Only the half-step dividing the major second remains to be accounted for.

Young considered different ways of assigning a value to this divisor of 9:8 in order to reproduce the auditory impression he had in mind. In the previously mentioned interview with Kostelanetz, Young suggests:

> In the most primitive form, I think of the ratios [in a Dream Chord] as 12, 16, 17, 18, which represent the intervals for *G, C, C♯, D.* However, in the version I gave them [Lukas Foss's performance group for a 1965 performance at Carnegie Recital Hall], I suggested 24, 32, 35, 36, because I was interested in the smaller interval of 35/36, which I felt was a ratio I may have been hearing all along.[22]

By the 1970s Young settled on the Dream Chord ratio of 12:16:17:18, which is used in all subsequent work and performances.

Trio for Strings

Of course, this immediately raises the possibility of placing some of Young's pre-*Tortoise* works in just intonation. For instance, *The Four Dreams of China* contains only the four notes of a Dream Chord heard in different voicings—so once the proportions of a Dream Chord are fixed, the tuning for the entire piece is immediately determined. Placing other early work in just intonation has been an ongoing project of Young's since the 1970s. He feels that just intonation more faithfully reproduces the auditory impressions he had in mind when writing the pieces and therefore is the appropriate context for the entirety of his work. Certainly, from our investigation of *Trio*, one can sense an affinity between the approaches.

However, there are more difficulties involved in placing Young's early twelve-tone pieces in just intonation. For one thing, all notes in the chromatic scale will necessarily appear. Furthermore, fixing a value of a note in the exposition does not guarantee that the note will have the same value throughout the piece. While row operations limit the intervallic relationships to those appearing in the row and their inversions, transposing and taking an inversion may force a note to stand in a new set of relations. On the other hand, in Young's twelve-tone pieces row operations were often selected for invariance, so many identical harmonic groupings from the exposition appear throughout the piece. In practice, providing a tuning for the exposition largely determines the tuning for the entire piece.

In 1978, Young devised a satisfactory tuning for his early piece *For Guitar*. This is a twelve-tone work, although in the spirit of experimentation with twelve-tone conventions, the row consists only of eleven notes. Shortly after completing this project, Young began searching for an appropriate tuning for *Trio*. Perhaps due to the significance of the piece, this ended up being a lengthy process. Following the approach in *For Guitar*, Young initially attempted to focus on intervals frequently occurring in his music. He writes:

> By 1982 I had notated two possible just tunings for the *Trio for Strings*. Tuning A was a Pythagorean tuning consisting of twelve pitches produced by a cycle of eleven ascending fifths (3:2 intervals). This Pythagorean tuning was inspired by the fact that the bowed stringed instruments are tuned in perfect fifths. Tuning B was a Septimal-Pythagorean tuning consisting of twelve pitches where ten produced a cycle of nine ascending perfect fifths (3:2 intervals) and two septimal intervals in the ratio of seven times the first partial (fundamental) and seven times the third partial. The two septimal intervals emphasized a blues element that exists in the *Trio*.[23]

However, these initial Pythagorean tunings (possibly based on Young's idea of the piece having an implicit tonal center of *C*) were deemed inadequate, and the project continued in fits and starts for years to come.

In the meantime, the piece was being performed in successively new configurations. In 1982, Young introduced a quartet version, with the addition of double bass to articulate some pitches as harmonics. In 1987, Young presented the piece at his thirty-year retrospective, working for the first time with the renowned cellist Charles Curtis, who became a consummate interpreter and performer of Young's work.

Subsequent to this event, Curtis worked closely with Young on the recent melodic versions of *The Four Dreams of China* and led performances of these pieces over the world. Curtis cites this experience as a key backdrop to *Trio*'s eventual tuning. Although the tuning for the *Dreams* are immediately given, since they are composed entirely of the four pitches in a Dream Chord, the process of constructing these intervals in performance by tuning to higher partials (in particular tuning the 17) gave Curtis a notion of applying this procedure backwards to *Trio*.

This process started with the fortieth anniversary of *Trio*, where Young and Curtis worked on reconstructing *The Original First Draft of The Exposition Section Alone* containing the exposition of the piece, but scaled to the length appearing in the originally notated sketches. Given these initial proportions, which were set aside at the time due to performance considerations, the row would take nearly forty minutes to complete, and the entire piece would stretch to nearly three hours. From his experience tuning intervals in *Four Dreams,* Curtis convinced Young that all double stops needed to be played by two players, one on each pitch, necessitating a new quartet version.

Trio and Tuning

Interestingly, the Pythagorean A and B tunings that Young cites above do not make any explicit use of *Trio*'s Dream Chord structure. Instead, note values are determined by an ascending series of fifths (with the Pythagorean B tuning containing the seventh partial of the fundamental and its second overtone.) In particular, neither of the Pythagorean tunings make use of the seventeenth partial of the fundamental, an intrinsic component of the piece. Through his experience performing the melodic versions of *Four Dreams*, Curtis began to get a feeling for how the four constellations in the exposition of *Trio* might be tuned to each other in performance.

Recall the exposition of *Trio* contains four pitch groupings, each of which are subsets of a Dream Chord, that can be expressed in prime form as:

$$\{\{C\sharp, D, E\flat\}, \{B, E, F, F\sharp\}, \{A\flat, A, B\flat\}, \{C, G\}\}$$

So locally each of the constellations can be assigned their appropriate Dream Chord values, as occurs in *Four Dreams*. For instance, {A♭, A, B♭} can be assigned the ratio 16:17:18.

Notice that this does not determine a unified tuning for the piece (or even the exposition), as the local values within these groupings are not related to each other in the context of a single overtone series. For instance, it is undetermined where the C♯ in the first constellation should be situated with respect to the C in the fourth. As it stands, there is no presumed fundamental to the piece, from which all other notes are related as partials. Instead, there are four separate, successively transposed, Dream Chords.

Given the intervals of silence separating the Dream Chords, it's not entirely unreasonable to view this local notion of tuning as sufficient. On the other hand, given Young's theory of Vertical Hearing (or Hearing in the Present Tense), where all intervals occurring in the piece are present at every moment as overtones of a fundamental, it also makes sense that these Dream Chords should be related to each other across the silence by their placement in a single overtone series. Even if the construction of each Dream Chord is an intrinsically local process, the chords should implicitly fit together as overtones of a single continuous fundamental. It is just a matter of hearing it in this way. To achieve this result, it's necessary to "tune" the constellations to each other.

Although the eventual tuning was not conceived in exactly this manner, an easy way to explain this idea is by choosing a "tuning note" from each of the four groupings. Then these four notes can themselves be assigned a ratio in such a way that preserves local Dream Chord structures. This creates a tuning for the exposition, and by extension the entire piece.

Of course, there are many ways of doing this. For instance, one could attempt a Pythagorean tuning like the ones considered above, where the tuning notes from each constellation form an ascending series of fifths. One choice is to select the tuning notes G, D, A, E from the four constellations and relate them by the ratio 8:12:18:27. Working out the details and removing all factors of two (since ascending or descending any number of octaves results in the same note) yields the tuning:

C: $3^1 \times 17$
G: $3^2 \times 17$
D: $3^3 \times 17$
A: $3^4 \times 17$
E: $3^5 \times 17$
B: $3^6 \times 17$
$F♯$: $3^3 \times 17$
$C♯$: 3^3
$A♭$: 3^4

$E\flat$: 3^5
$B\flat$: 3^6
F: $3^5 \times 17^2$

This is a global or unified tuning for the exposition, and by extension the entire piece, in the sense that all notes are now related to each other as overtones of a single fundamental. However, like Young's initial Pythagorean tunings, there are some drawbacks to this proposal. For instance, the fundamental (the note corresponding to the value 3^0) does not appear in the exposition. Working backwards by fifths from $C\sharp$ we see that the fundamental should correspond to the note E, which is actually represented by the value $3^5 \times 17$. The strong disparity between the presumed fundamental and its actual value in the tuning is not appealing.

Furthermore, while each of the four constellations are tuned to each other, they are not aligned in any especially revealing way. A natural tuning would give the sense of excerpting from a larger and potentially infinite structure, by carving out precisely defined cross sections of the harmonic series. In the present situation, this would involve regular series of the third and seventeenth partials of the fundamental (since modulo two, the only prime factors of twelve, sixteen, seventeen, and eighteen are three and seventeen.)

In the above tuning, the highlighted partials do not reveal any obviously compelling structure. The two series based on the third and seventeenth partials do not start at the beginning or work upwards in a regular way. Notice that each note in the full Dream Chord $\{B, E, F, F\sharp\}$ contains a factor of seventeen. As a result, the seventeen value in this constellation (F: $3^5 \times 17^2$) needs an additional factor of seventeen to ensure its correct local Dream Chord value. In the context of a global tuning, this situates F awkwardly in the series based on the third and seventeenth partials. While many unified tunings exist, there is still the question of finding a natural and performable one.

A Dream Chord Tuning

One might say the issues encountered with Pythagorean tunings arise because they are imposed on the piece by external considerations. Although a Dream Chord does contain the interval of a fifth, this method of relating notes by fifths is too coarse to capture the underlying harmonic structure of the piece.

Indeed, the eventual tuning did not arise from such extrinsic motivations, but rather from natural and explicit performance considerations woven into the piece itself. In the early 2000s, Charles Curtis, working backwards from his long experience tuning notes (sometimes over intervals of silence) in the melodic versions of *Four Dreams*, began to consider how a just intonation performance of *Trio* might work. He recollects:

The main breakthrough in finding a tuning was inseparable from my approach as a cellist, looking from the inside of performing for a way to maintain a continuity of tuning *by ear*—not with a sine wave in headphones or a digital tuner to look at—across silences and from transposition to transposition. . . . So the first impulse was to ask myself, if I were to tune one constellation to the next, what notes would I focus on, which ones could I really hear as pure intervals . . . [24]

His idea of a unified tuning was one manifested through performance, rather than through the application of abstract external principles. Of course, this closely accords with Young's general compositional philosophy.

Looking at how the Dream Chords are successively transposed, Curtis considered tuning notes from each constellation that formed intervals familiar in the piece. In fact, he began to consider the most familiar interval of all: the Dream Chord itself. Indeed, looking at the four constellations, we see that choosing the notes C♯, F♯, A♭ and G from each successive constellation yields a Dream Chord that can be expressed in prime form as C♯, F♯, G, A♭. Interestingly (by a combinatorial argument I expressed to Curtis at the time), one can show this is the *unique* choice of tuning notes that form a Dream Chord. Hence, there is a single way to tune the constellations to each other through this hidden Dream Chord structure, which is always present but never explicitly heard.

Although Curtis conceived this tuning in terms of performance, it is certainly instructive to work out how the notes are situated together in a single overtone series. As mentioned above, the local values within the constellations are determined by their Dream Chord structure, and additionally we want the notes C♯, F♯, G, A♭ to have the ratio 12:16:17:18. The only issue, which is easily dealt with, is maintaining the desired local ratios when using the tuning notes to relate the constellations.

For instance, consider the first two constellations $\{C♯, D, E♭\}$ and $\{B, E, F, F♯\}$. The tuning notes C♯ and F♯ have local values 16 and 18 in their respective groupings, and globally they need to be related by the ratio 12:16. Clearly multiplying every note in a constellation by the same number will not affect local values.

To ensure the tuning notes (and consequently all notes in the first two constellations) have the correct global values, it suffices to multiply the value of every note in the first constellation by 18×12 and the value of every note in the second constellation by 16×16. Then by construction C♯ and F♯ will have the global ratio $16 \times (18 \times 12) : 18 \times (16 \times 16) = 12:16$, and also have the local values of 16 and 18 within their groupings. Continuing this basic process yields the unified tuning:

C♯	$16 \times (18 \times 12)$
D	$17 \times (18 \times 12)$
E♭	$18 \times (18 \times 12)$
B	$12 \times (16 \times 16)$
E	$16 \times (16 \times 16)$
F	$17 \times (16 \times 16)$
F♯	$18 \times (16 \times 16)$
A♭	$16 \times (18 \times 18)$
A	$17 \times (18 \times 18)$
B♭	$17 \times (18 \times 18)$
C	$18 \times (18 \times 18)$
G	$18 \times (18 \times 17)$

In this way, all notes are related to each other in terms of a single overtone series. To explicitly view this series, it is better to express the notes in terms of their prime factorization (modulo octave equivalence):

E	3^0
B	3^1
F♯	3^2
C♯	3^3
A♭	3^4
A♭	3^5
B♭	3^6
F	$3^0 \times 17$
C	$3^1 \times 17$
G	$3^2 \times 17$
D	$3^3 \times 17$
A	$3^4 \times 17$

This tuning illuminates two nicely defined and natural ascending series based on the third and seventeenth partials, which structurally serve as the irreducible harmonic building blocks of the piece. One series starts from the fundamental E and ascends by fifths. The other begins at the seventeenth partial, also ascending by a series of fifths. This relationship between the third and seventeenth partials highlights another aspect of just intonation appealing to Young: not only does it clarify existing harmonic language, but it explores harmony beyond the standard Western tradition. Many of his pieces make explicit use of harmonic content that had been perceived previously only implicitly as overtones of a fundamental. In the present case, given that the ratio 17:18 is supposed to represent a minor second, one might observe that while the interval itself is not unfamiliar, when performed in just intonation through tuning to partials it does possess a noticeably different auditory quality. Indeed, this is one of the virtues of this new version of *Trio*. Although traditional harmony is derived from the overtones series, just intonation offers an expanded theory, both clarifying existing harmonic language and exploring at times unfamiliar regions of consonance within a unified framework.

In the summer of 2015, nearly sixty years after its initial conception, the first performance of *Trio* in its originally conceived duration and with the Dream Chord tuning took place at Dia Chelsea. This performance, led by Curtis, became the basis for the first commercial release of the piece by Dia in 2022. The ongoing development of *Trio* reflects not only the history of minimalism from its earliest stages, but also serves as an interesting reflection of Young's compositional process taken as a whole. In this setting, works are not treated as completed objects, but serve as generative elements and continued constructions, developing in time and always subject to further refinement. Indeed, this compositional model, witnessed in *Trio*'s evolving performance history and eventual just intonation tuning, is developed explicitly and formally in later pieces such as the *Melodic Versions* of *The Four Dreams of China*. Furthermore, this conception closely mirrors Young's key theoretical constructs, such as tuning as a function of time and Vertical Hearing (or Hearing in the Present Tense), which emerge from primitive notions of continuity and constructability. While the novel ideas surrounding *Trio*'s creation have remained constant, and its style and sound look prescient for much of what followed in the 1960s, it has taken another sixty years for the appropriate context to coalesce, enabling the piece to be appreciated more fully and in greater detail than ever before.

1   La Monte Young, "Leonard Stein (1916–2004): A Memorial Tribute," in *Trio for Strings* (New York: Dia, 2022), 15.

2   Jeremy Grimshaw, *Draw a Straight Line and Follow It* (London: Oxford University Press, 2012), 198.

3   Although not totally standard, to highlight the group-theoretic interpretation we choose to include the identity permutation $T=T_0$.

4   Arnold Schoenberg, "Composition with Twelve Tones," in *Style and Idea* (New York: Philosophical Library, 1950), 113–14.

5   Schoenberg, "Composition with Twelve Tones," 109.

6   Schoenberg, "Composition with Twelve Tones," 109.

7   Schoenberg, "Composition with Twelve Tones," 110.

8   Schoenberg, "Composition with Twelve Tones," 109.

9   Schoenberg, "Composition with Twelve Tones," 109.

10   Dorothy Lamb Crawford, "Arnold Schoenberg in Los Angeles," *The Musical Quarterly* 86, no. 1 (2002): 6–48.

11   Schoenberg, "Composition with Twelve Tones," 113–14.

12   Schoenberg, "Composition with Twelve Tones," 109.

13   La Monte Young, "Notes on Trio for Strings," in *Trio for Strings* (New York: Dia, 2021), 9.

14   Richard Kostelanetz, "Conversation with La Monte Young," in *Selected Writings*, ed. La Monte Young and Marian Zazeela (ubuclassics, 2004), 23.

15   Kostelanetz, "Conversation with La Monte Young," 47.

16   Keith Potter, *Four Musical Minimalists: La Monte Young, Terry Riley, Steve Reich, Philip Glass* (Cambridge: Cambridge University Press, 2002), 39.

17   Kostelanetz, "Conversation with La Monte Young," 26.

18   Young, "Notes on Trio for Strings," 8.

19   La Monte Young, statement to author, July 2006.

20   La Monte Young and Marian Zazeela, "Dream House," in La Monte Young and Marian Zazeela, eds., *Selected Writings* (ubuclassics, 2004), 15.

21   Young and Zazeela, "Dream House," 12.

22   Kostelanetz, "Conversation with La Monte Young," 47.

23   Young, "Notes on Trio for Strings," 13.

24   Charles Curtis, correspondence with author, April 2024.

Ticking Stripe by Spencer Gerhardt

Published by Blank Forms Editions
Artistic Director   Lawrence Kumpf

Copyediting   Ciarán Finlayson and Dana Kopel
Proofreading   Heather Holmes
Design   Alec Mapes-Frances

Printed by Ofset Yapımevi in Turkey

"Introduction to Matt Connors: Gui(l)de" originally published in M. Connors, *Gui(l)de* (Karma, 2019)

"Topological Generation of Special Linear Groups" originally published in *Journal of Algebra* 569, 2021

"Reverse Perspectives" originally published in *Charles Curtis: Performances and Recordings 1998–2018* (Saltern, 2020)

"Minimalism and Foundations" originally published in R. Kossak, P. Ording, eds., *Simplicity: Ideals of Practice in Mathematics and the Arts, Mathematics, Culture, and the Arts* (Springer International Publishing AG, 2017)

"Order Three Normalizers of 2-Groups" originally published in *Communications in Algebra* 45 (Taylor & Francis Ltd, 2017) and reprinted by permission of the publisher. http://www.tandfonline.com

"Domains of Variation" originally published in *Intelligent Life* (Blank Forms Editions, 2019)

Blank Forms Editions is supported by the Robert Rauschenberg Foundation, The Andy Warhol Foundation for the Visual Arts, Agnes Gund, and the Blank Forms Publisher's Circle, including Desiree and Olivier Berggruen, Steven Eckler, Gisela Gamper, Abigail Goodman and David Norr, Linden Renz, and Mary Wang and Christian Nyampeta.

ISBN 978-1-953691-21-7

Blank Forms
468 Grand Avenue
#1D/#3D
Brooklyn, NY 11238
blankforms.org